An Introduction to Space Weather

Our space age technology enables global communication, navigation, and power distribution that have given rise to our "smart," interconnected, and spacefaring world. Much of the infrastructure modern society depends on, to live on Earth and to explore space, is susceptible to space weather storms originating from the Sun. The second edition of this introductory textbook is expanded to reflect our increased understanding from more than a dozen scientific missions over the past decade. Updates include discussions of the rapidly expanding commercial space sector, orbital debris and collision hazards, our understanding of solar–terrestrial connections to climate, and the renewed emphasis on human exploration of the Moon and Mars. The book provides new learning features to help students understand the science and solve meaningful problems, including some based on real-world data. Each chapter includes learning objectives and supplements that provide descriptions of the science, and learning strategies to help students and instructors alike.

MARK MOLDWIN is an Arthur F. Thurnau Professor of Climate and Space Sciences and Engineering at the University of Michigan, Ann Arbor. An award-winning teacher, his primary research interests are magnetospheric, ionospheric, and heliospheric plasma physics, and pre-college space science education and outreach. He is a Past-President of the American Geophysical Union's Education Section, a former Editor-in-Chief of *Reviews of Geophysics*, and among his numerous honors he received the 2016 AGU's Waldo E. Smith Award for his "extraordinary service to geophysics."

"Professor Moldwin lays out the complex topic of space weather in a clear and comprehensive way, with a masterful educational touch that keeps the reader engaged in every chapter. The second edition of this foundational book comes at the perfect time, as the space economy is rapidly expanding and the topic of space weather becomes more pertinent than ever before."

Professor Jacob Bortnik, University of California, Los Angeles

"From solar explosions to their impact on society, this wonderful text uses understandable prose to introduce students with widely different backgrounds to the physics of the local universe and to how the ever-changing connections between Sun and Earth affect society. For the second edition, the diverse supplemental material – ranging from how science works to the foundations of physics – has been expanded, while a newly developed companion webpage leads to beautiful images, illustrative videos, and ready-to-use lecture slide sets."

Dr. Karel Schrijver, Lockheed Martin Advanced Technology Center

"A very fine introduction to the science and applications of the broad topic of solar–terrestrial physics (space weather) by Professor Moldwin, an eminent scholar who has personally contributed substantially to the field. In this substantially revised edition, the clever division of each chapter into descriptive and more quantitative portions will make this volume the preeminent academic introduction to the subject."

Professor Louis J. Lanzerotti, New Jersey Institute of Technology

"We have used the first edition of *An Introduction to Space Weather* for several years in Penn State's cross-listed 'Space Weather' course; it reaches our students well, regardless of their diverse backgrounds. This new edition brings welcome updates (including a brand new chapter on forecasting, educational and other supplementary content toward the end of each section, as well as a companion website) while remaining concise and affordable."

Professor Timothy Kane, The Pennsylvania State University

"I have used Professor Moldwin's book in Embry–Riddle's second-year undergraduate class 'Introduction to Space Weather'. Professor Moldwin's book has worked as a great and highly motivating intro-ductory text for important space weather phenomena. It has been developed with pedagogy and student learning in mind. The book is well written, clear, and captivating. The freshmen students liked the conceptual nature of the text, so the content was easy to grasp with only some introductory physics/math background. I plan on continuing to use this book when teaching this class next time."

Professor Heidi Nykyri, Embry–Riddle Aeronautical University

An Introduction
to Space Weather

SECOND EDITION

Mark Moldwin
University of Michigan,
Ann Arbor

CAMBRIDGE
UNIVERSITY PRESS

Shaftesbury Road, Cambridge CB2 8EA, United Kingdom

One Liberty Plaza, 20th Floor, New York, NY 10006, USA

477 Williamstown Road, Port Melbourne, VIC 3207, Australia

314–321, 3rd Floor, Plot 3, Splendor Forum, Jasola District Centre,
New Delhi – 110025, India

103 Penang Road, #05–06/07, Visioncrest Commercial, Singapore 238467

Cambridge University Press is part of Cambridge University Press & Assessment,
a department of the University of Cambridge.

We share the University's mission to contribute to society through the pursuit of
education, learning, and research at the highest international levels of excellence.

www.cambridge.org
Information on this title: www.cambridge.org/highereducation/isbn/9781108791717
DOI: 10.1017/9781108866538

First published 2008
Second edition 2023

Printed in the United Kingdom by TJ Books Limited, Padstow, Cornwall

A catalogue record for this publication is available from the British Library.

ISBN 978-1-108-79171-7 Paperback

Additional resources for this publication at www.cambridge.org/moldwin2.

Contents

The color plate section is located between
pages 52 and 53.

Preface

In the last few decades our technological civilization has become dependent on satellites for global communication, navigation, and commerce. We have also begun the long journey to explore the Moon, Mars, and our solar system.

This exploration has led to some amazing discoveries about our dynamic Sun and its interaction with Earth. We now know that the Sun is a variable star that expels high-energy particles and deadly radiation continuously out into space. This radiation can impact and destroy technological systems and is one of the major concerns for modern civilization and human space exploration.

In the 1990s, the commercial satellite industry boomed with direct satellite-to-home TV and radio markets and satellite communication options expanding. By the 2000s, the satellite industry was doing over $100 billion (US) per year of business with nearly a hundred new satellites launched each year. In 2020, global satellite revenue was over $370 billion. With the increasing number of commercial space businesses and the reliance that different markets started to have on space, society began to notice when something went wrong in space.

Galaxy IV was an operating and profitable communications satellite until May 19, 1998, when, after experiencing weeks of intense radiation generated by the Sun and the Sun's interaction with the Earth's space environment, it failed. Galaxy IV carried the signals of over 90% of North America's pagers (early wireless communication devices) and several major broadcast networks including the US National Public Radio (NPR) and CBS. Without the $200 million satellite, millions of pager messages and NPR radio and CBS television programs never made it to their intended audience. Radio and TV producers were left scrambling to fill dead-air time, and medical doctors and business people found themselves out of contact with their hospitals and clients. Galaxy IV in all likelihood was a victim of a space weather storm.

Though the cause of failure is not yet known, SiriusXM's geosynchronous SXM-7 satellite failed while on-orbit in early 2021 just six weeks after successful launch, becoming a $225 million "total loss." Even if the cause of failure was not space weather related, the "dead" satellite needed to put into a graveyard orbit, otherwise it would have been a space weather hazard for other geosynchronous spacecraft due to potential collisions. The launch of more and more satellites increases the number of orbiting objects, which increases the number of operational spacecraft, derelict spacecraft, and orbital debris that become space weather collision hazards.

Space storms can not only damage or destroy orbiting satellites, but also injure or kill astronauts, degrade or blackout certain radio and navigation communications, and cause regional power failures by destroying critical components of electrical power grids. With the continued growth of the satellite communications industry, the advent of space tourism, and our growing dependence on wireless communication and instant access to global information, we are becoming more and more susceptible to problems caused by space weather.

This textbook aims to introduce the reader to the field of space weather in both a descriptive and quantitative approach. The mathematical sophistication of the reader is assumed to be at the level of high school algebra. Since science is not just a collection of facts, but a process or way of understanding our natural world, the book attempts to answer the question "How do we know that?" by including discussions on the historical development of different concepts.

The first edition of this book was derived from the notes for three undergraduate courses at UCLA – the first a freshman seminar, the second an Honors Collegium course, and the third a general education course for non-science majors entitled "The Perils of Space: An Introduction to Space Weather," first taught in Fall 2004. The second edition of this book benefits from SPACE103 Introduction to Space Weather, first taught at the University of Michigan in 2014.

Each chapter is divided into two separate parts: the main text describing space weather topics, and supplements describing important physical concepts behind each topic. In this second edition, the supplements also include discussions on learning theory and study skills that are aimed to help students develop their conceptual understanding. End-of-chapter problems allow students to test their understanding and delve deeper into aspects of the chapter. A list of key concepts is given for each chapter, and the concepts are in bold at their first occurrence in the main body of the text. Finally, learning objectives follow the list of key concepts at the beginning of each chapter, describing what the reader should be able to *do* after reading the chapter. Additional learning objectives at the start of the supplements to each chapter describe what students should be able to do after reading the supplements. Students wishing to understand space weather should conceptually know the key concepts and be able to use that information to analyze, synthesize, and evaluate a variety of space weather problems and questions. Key concepts are indicated by bold in the index. The second edition now also includes a glossary of important terms and key concepts to provide understanding of overarching processes, regions, events, and the connections between them.

A new aspect of the second edition are the online resources available at www.cambridge.org/moldwin2 that provide access to data of the Sun, solar wind, the Earth's magnetosphere, and upper atmosphere that can be used to practice applying space weather concepts to actual observations. (For example: Can you identify a coronal mass ejection [CME] in solar coronagraph images and its signatures in the solar wind? With that identification, can you estimate the arrival time of the disturbances at Earth, and what will that look like in different regions of the Earth's space environment?) The online resources also include a variety of lecture slides, images, and videos that help visualize the key regions, processes, and events outlined in the text.

Acknowledgments

Students in the "Introduction to Space Weather" courses at the University of Michigan inspired and helped in the writing of the second edition of this textbook. Several colleagues who have adopted the text at their university have provided ideas, suggestions, and support, and I owe them thanks and gratitude for helping to spread awareness of space weather more broadly. Much of the revision of the textbook took place during a sabbatical supported by the US–Norway Fulbright Arctic Chair visit to the University of Bergen's Birkeland Center for Space Science. My hosts (Professor Michael Hesse, Dr. Therese Moretto Jorgensen, Professor Nikolai Østgaard, and Kavitha Østgaard) provided office space and resources, and welcomed me into their research groups, for which I am very grateful. I also would like to thank my University of Michigan colleagues and my Magnetometer Lab science, engineering, and art students for their support and motivation during the years of writing this revised edition of the book.

Chapter 1
What is Space Weather?

"Space weather" refers to conditions on the sun and in the solar wind, magnetosphere, ionosphere, and thermosphere that can influence the performance and reliability of space-borne and ground-based technological systems and can endanger human life or health. Adverse conditions in the space environment can cause disruption of satellite operations, communications, navigation, and electric power distribution grids, leading to a variety of socioeconomic losses.

(National Space Weather Program, 1995)

1.1 Key Concepts

- space weather
- climate
- meteorology
- Earth's atmosphere
- Systems Science

1.2 Learning Objectives

Recommendation: read Section 1.6.7 before reading this chapter to learn and utilize active reading techniques.

> After actively reading this chapter, readers will be able to:
>
> - describe how the Sun influences the Earth's space environment and discuss the connection between storms on the Sun, the aurorae and geomagnetic disturbances on Earth;
> - summarize the history of space weather in the context of early observations of the Sun, the geomagnetic field, and the northern lights;
> - develop a list of components of the space system and determine how modern society is impacted by space weather.

1.3 Introduction

Since the 1960s, we have become a spacefaring civilization. With robotic and manned spacecraft, we have started to survey our solar system. We have learned that we live in the atmosphere of a dynamic, violent Sun that provides energy for life on Earth but can also cause havoc among its fleet of satellite and communications systems. **Space weather** is the emerging field within the space sciences that studies how the Sun influences Earth's space environment and the technological and societal impacts of that interaction: damage to or destruction of Earth-orbiting satellites, threats to both astronaut safety during long-duration missions to the Moon and Mars, and threats to the reliability and accuracy of global communications and navigation systems.

Modern society depends on accurate forecasts of weather (day-to-day variability of temperature, humidity, rain, etc.) and understanding of **climate** (long-term weather trends) for commerce, agriculture, transportation, energy policy, and natural disaster mitigation. The science of understanding weather, **meteorology**, is one of the oldest human endeavors to make sense of our natural environment. Like meteorology, space weather seeks to understand and predict climate and weather, but of outer space. For millennia, space storms have raged above our heads unknown to us. But with the advent of the space age, we have begun to notice the destructive power of severe space weather.

Like weather, space weather has its roots in the Sun. The main distinctions between the two types of weather are where it takes place and the type of energy from the Sun that influences it. For weather, we are most concerned with the troposphere, which extends from Earth's surface to the top of the highest clouds at about 10 km. Space weather is interested in the space environment around Earth all the way to the Sun. Space begins in a region of **Earth's atmosphere** called the thermosphere, which starts at an altitude of roughly 100 km. The International Space Station flies at an altitude of about 350 km. Plate 1 shows a picture of Earth's atmosphere from the International Space Station. The sharp contrast between the blue of Earth's atmosphere and the blackness of space is at approximately 100 km.

The second difference between weather and space weather is the type of solar energy that influences the two regions. The Sun continuously emits two main types of energy into space – electromagnetic (EM) radiation and corpuscular radiation. Visible light, radio waves, microwaves, infrared, ultraviolet (UV), X-rays, and gamma rays are forms of EM radiation. The Sun's EM radiation bathes the top of Earth's atmosphere with about 1400 watts[1] of power per square meter and heats the

---

[1]   A watt is the SI unit of power (energy per time) named in honor of James Watt (1736–1819), a Scottish engineer and scientist credited with making the steam engine a practical device.

lower atmosphere, surface, and oceans unevenly. Winds are driven by these differences in atmospheric temperature.

The Sun also continuously emits corpuscular (minute particle) radiation, charged atoms, and subatomic particles (mostly protons and electrons) in what is called the solar wind. Like winds on Earth, the solar wind is driven by temperature differences, but those differences are between the Sun's upper atmosphere and interplanetary space. The solar wind, which expands out into the solar system carrying with it the Sun's magnetic field, carves out a region of interstellar space called the heliosphere (from "helios," Greek for "Sun").

The solar wind is not steady or uniform but changes constantly. These changes affect Earth's space environment in a number of ways, including the creation of new corpuscular radiation that bombards Earth's upper atmosphere, causing aurorae (northern and southern lights) and large electrical currents that can disrupt communication, power grids, and satellite navigation.

Occasionally the Sun's surface erupts and sends a large part of the solar atmosphere streaming away at high speeds. These events, called coronal mass ejections (CMEs), can contain 10^{12} kg (or 1 000 000 000 000 kg) of material (equivalent to a quarter of a million aircraft carriers) and can move away from the Sun at over 1000 km s^{-1} (over several million miles per hour) (Plate 2). If CMEs are directed towards Earth, a great space storm can develop far above our heads, crippling satellites, causing increased radiation exposure for airline crews and passengers, blacking out some forms of radio communication, and disrupting power systems on Earth.

These space storms, like weather storms such as Hurricane Sandy in 2012, have caused severe damage to technological systems in the past. In March 1989, a large CME slammed into Earth causing massive power outages in eastern Canada. The US Government considers space weather to be one of the most serious natural hazards facing the nation's critical infrastructure. It is one of two natural hazards described that can impact the entire nation's economy – the other is a global pandemic. The emerging science of space weather is attempting to understand the causes of space storms and their impact on Earth's technological infrastructure with the hope that we can forecast space weather and mitigate damage.

1.3.1 It's Greek to Me: the Origin of Technical Names in Science

The ancient Greeks envisioned the heavens as being on concentric spheres around Earth, with the planets (Greek for "wanderers"), Sun,

Table 1.1 *Greek prefixes for regions of Earth's atmosphere*

Prefix	English translation	Height	Characteristic
tropo	mixing or changing	0–10 km	where weather takes place
strato	layer	10–50 km	where ozone layer is located
meso	middle	50–80 km	coldest region
thermo	heat	~1000 km	where space begins
iono	to go	~1000 km	where aurorae occur

and Moon moving on their own celestial spheres, while the stars moved in lock-step behind them on their own sphere. Science borrows from this worldview by giving concentric regions in and around the planets and the Sun a Greek prefix and the suffix "sphere." The rocky surface of Earth is often called the lithosphere ("litho," meaning "stone"), the water part the hydrosphere ("hydro," meaning "water"), the place where life is found the biosphere ("bio," meaning "life"). The region above Earth's surface is called the atmosphere ("atmos," meaning "vapors"). The atmosphere is further divided into subregions, which are listed in Table 1.1. The boundaries between the spheres are called "pauses" (e.g., the boundary between the troposphere and stratosphere is the tropopause). Several more "spheres" and "pauses" will be introduced in the following chapters. Figure 1.1 shows the layers of the atmosphere as a function of height. Note that each layer or sphere has a different temperature profile with height. For example, the troposphere has temperature decreasing with altitude, while the stratosphere has temperature increasing with altitude.

1.4 Brief History

The study of space weather began with systematic observation of three natural phenomena: the aurorae (also called the northern or southern lights), Earth's magnetic field, and sunspots (dark regions observed on the surface of the Sun). Because aurorae can be seen with the unaided eye, they have been observed for thousands of years, though the systematic study of the aurorae didn't begin until the sixteenth century. Development of the sensitive compass and telescope in the early seventeenth century made possible the discovery of the nature of Earth's magnetic field and sunspots.

Understanding of space weather traces its roots to the connection of these three phenomena. The first tentative connections were made in the middle of the nineteenth century. Since then, we have slowly expanded our

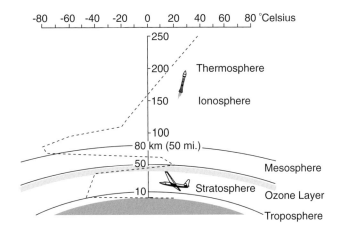

Figure 1.1 The vertical temperature scale of Earth's atmosphere. The dashed line represents the temperature as a function of height. Each region is defined by how the temperature changes with height. (Courtesy of Cislunar Aerospace, Inc.)

knowledge of the Sun and Earth's space environments and, in so doing, have begun to develop a physical model of the Sun–Earth connection. This section gives a brief history of the discoveries and an introduction to some of the scientists who have led us to our current understanding of the solar–terrestrial relationship. As with all areas of science, the field of space weather developed in concert with our understanding of physics and chemistry and new technologies that allowed us to "see" the "invisible" – things too small or far away to be seen with the unaided eye or beyond our sensibilities and capabilities to see, hear, or feel, such as radio waves and magnetic fields. A detailed timeline of our understanding of space weather is given in Appendix 4 and a timeline website that connects to the original scientific literature is also available. The web address is given in Appendix 5.

1.4.1 The Aurorae

Our earliest ancestors observed the aurorae. Until the eighteenth century most treatises on aurorae were based on speculation about their origin by men who may have never observed them. These speculations usually followed Aristotle's (384–322 BC) view of aurorae as burning flames, or Rene Descartes' (1596–1650) idea that aurorae were moonlight or sunlight reflected off ice or snow crystals. Systematic observations of aurorae were first made in the sixteenth century. One of the greatest astronomers of all time, Tycho Brahe (1546–1601), recorded the occurrence of aurorae between 1582 and 1598 from his Uraniborg observatory in Denmark. He found that the number of aurorae varies from year to year, but did not note any systematic or regular variation.

On September 12, 1621, the astronomer Pierre Gassendi (1592–1655) from the south of France and Galileo[2] in Venice observed the same aurorae. Gassendi called the lights aurora borealis (Latin for northern dawn), a name that has been associated with polar lights ever since. He noted that the aurorae must occur high in Earth's atmosphere for observers at distant locations to be able to observe the same phenomena.

In the eighteenth century a number of observations began to illuminate the origins of aurorae. Frenchman Jean-Jacques d'Ortour de Mairan (1678–1771) made the first rough measurements of auroral height in 1726; these were consistent with Gassendi's observation that aurorae occur in the upper atmosphere. Using the triangulation method, English scientist Henry Cavendish (1731–1810) correctly estimated auroral height to be between 80 and 112 km in 1790. However, estimates of auroral height continued to have large uncertainties until around 1900, when Norwegian scientist Carl Størmer (1874–1957), measured the height accurately using photographic techniques. Captain James Cook was the first European to observe the southern lights (which he called aurora australis) while in the Indian Ocean near latitude 58° S on February 17, 1773. He wrote in his ship's log:

> lights were seen in the heavens, similar to those in the northern
> hemisphere, known by the name of Aurora Borealis

In the nineteenth century as reports from polar explorers were compiled, it became clear that aurorae appear in large ovals centered near the North and South Poles. Captain John Franklin, who later perished with his crew as they attempted to find the Northwest Passage, determined that the number of auroral sightings decreases nearer the pole, suggesting an auroral zone. In 1833, the German geographer Georg Wilhelm Muncke (1772–1847) noted the existence of a zone of maximum auroral occurrence that is limited in latitude. In 1860, Professor Elias Loomis (1811–1888) of Yale University published the first map of the north polar region showing the zone where aurora had been most commonly observed (see Figure 1.2).

So, by the middle of the nineteenth century a number of facts about aurorae were known: they occur in an oval around the north and south

---

[2] Galileo Galilei (1564–1643), Italian physicist and astronomer, and founder of the modern scientific method. The first to use a telescope for astronomical observations, he discovered the four largest moons of Jupiter (named the Galilean moons in his honor); that Venus has phases, which offered direct support for the Copernican heliocentric theory; that the Milky Way was made up of individual stars; and that the Moon has mountains.

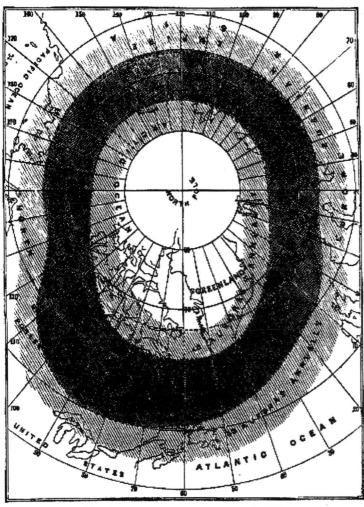

Fig. 18.—GEOGRAPHICAL DISTRIBUTION OF AURORAS.

Figure 1.2 The auroral oval from Professor Loomis' late-nineteenth-century study. Note that the aurora has a zone of occurrence centered around, but not at, the pole. (Source: Loomis, 1869)

polar regions, and they are high in the upper atmosphere. The search was still on for the cause of the aurorae.

1.4.2 The Geomagnetic Field

In 1088, Chinese encyclopedist Shen Kua (1031–1095) wrote the first description of compasses, magnetized pins floated on a small cork in a bowl of water. Alexander Neckham of St. Albans (1157–1217) was the first European to describe a compass in his work *On the Nature of*

Things published in 1187. Neckham had probably heard of the Chinese compass through the Silk Road trade routes from China to Western Europe. In 1576, Robert Norman discovered that Earth's magnetic field has a vertical component called dip. Combining this discovery with his own work on a model magnetic field called a terrella, William Gilbert[3] (later the personal physician of Queen Elizabeth I) wrote a book called *De Magnete* in 1600. In this book he demonstrated that Earth's magnetic field behaves like a magnet, which led to the systematic study of magnetic field orientation as a function of position on Earth. These magnetic maps allowed the use of compasses for navigation. In 1722, George Graham (1674?–1751) built a compass sensitive enough to observe slight (usually less than 1°) irregular variations of the geomagnetic field that caused the compass needle to "wiggle" slightly.

Thus the basic facts about geomagnetism known by the early eighteenth century were that Earth has a magnetic field like that of a refrigerator magnet (called a dipole magnetic field) with both regular and irregular variations. The search was on for the cause of these geomagnetic fluctuations.

1.4.3 Sunspots

In 1610, Galileo turned his telescope to the Sun and observed sunspots by focusing the image onto a piece of paper (see Figure 1.3). Several other observers – Johannes Fabricius, Thomas Harriot, and Christoph Scheiner – essentially simultaneously observed sunspots with the newly developed telescope. But what were these? Scheiner argued that they were moons or planets (Mercury, Venus, or the mythical planet Vulcan) orbiting between the Sun and Earth, while Galileo argued that they were on the surface of the Sun. The first regular daily observations (when weather permitted) of sunspots began in 1749 at the Zurich Observatory in Switzerland. Using data from Zurich, Samuel Heinrich Schwabe (1789–1875) recognized the occurrence of an 11-year solar cycle in about 1844. His original goal had been to find intra-mercurial planets such as those conjectured in Galileo's time. He began to systematically look for "transits" of these hypothesized planets across the Sun. In so doing, he meticulously recorded the position of every sunspot for 18 years. With this data set he discovered the 11-year sunspot cycle. He never did discover a planet crossing the Sun.

---

[3] William Gilbert (1544–1603), English physicist and physician who pioneered the field of geomagnetism. The first English scientist to accept the Copernican view of the solar system, he suggested that habitable worlds might be in orbit around other stars.

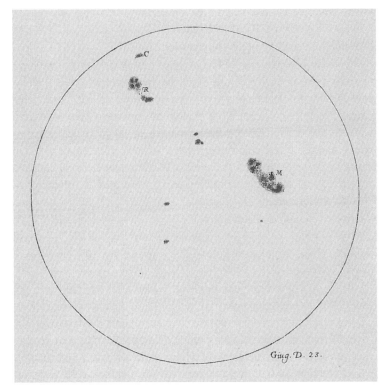

Figure 1.3 Sunspot drawings from Galileo made in about 1610. (Source: Galileo Galilei, 1613, courtesy of Owen Gingerich)

So, by the mid nineteenth century, it was clearly demonstrated that sunspots exist on the Sun and they have an 11-year cycle during which their number waxes and wanes. The question of what sunspots are and what, if any, effect they have on Earth remained.

1.4.4 Making the Connection between Aurorae, the Geomagnetic Field, and Sunspots

English astronomer Edmond Halley (1656–1742), of Halley's Comet fame, noted that aurorae that occurred on March 16, 1716, over London appeared to have the rays converging toward Earth similar to Earth's magnetic field lines. He drew the magnetic field lines outside Earth by extrapolating from the shape that iron filings make around a magnet. This hint that auroral rays are aligned with Earth's magnetic field was not confirmed until 1770 when the Swedish scientist Johann Wilke observed that auroral rays actually lie along the geomagnetic field lines. In 1722, the British instrument maker George Graham observed slight magnetic fluctuations with his compass, which later were observed to be correlated with observations of aurorae by Anders Celsius (the scientist whose name was given to a temperature scale) and his student (and brother-in-law) Olaf Hiorter in Uppsala, Sweden, in 1747. Professor Celsius' instrument was obtained from Graham, and through regular correspondence Celsius and Graham found that days with geomagnetic activity (the name given to magnetic fluctuations) in London were also days with geomagnetic activity in Uppsala. This established that geomagnetic activity and hence aurorae occur over large distances. Hiorter wrote "That aurorae must be the highest phenomena of our atmosphere, so high and extensive, that they can simultaneously, here and in England, at Uppsala and London, ... disturb the magnetic needle." They and others then began to observe periods of very large geomagnetic fluctuations – many degrees of magnetic needle fluctuation in several minutes. These large geomagnetic disturbances are now called geomagnetic storms.

In the mid nineteenth century, Col. Edward Sabine and Professor Rudolf Wolf were independently the first to publish the results showing the correlation between sunspots and geomagnetic activity. Sabine was a British military officer (later knighted) in charge of British observatories around the world at which surface weather and changes in the geomagnetic field were monitored. Understanding the geomagnetic field was important for navigation because of the use of magnetic compasses. Rudolf Wolf was director of the Zurich Observatory and therefore had access to the longest record of sunspot occurrence in the world. The studies of Graham and Celsius now linked aurorae with geomagnetic activity, and the studies of Sabine and Wolf linked geomagnetic activity with solar activity. Because of the unreliable reporting of the occurrence of aurorae (summer time, clouds, limited populations at high latitudes), it was several years before the occurrence of aurorae was also clearly shown to be linked with solar activity.

Immediately the question was raised asking what connected sunspots with Earth's magnetic field? In 1859 when Richard Carrington, who worked as an astronomer at Greenwich Observatory, observed a white light flare above a sunspot, he remarked in a letter to the British Royal Society that the flare was followed within a day by a huge geomagnetic disturbance. (Figure 1.4 shows Carrington's drawing of the flare.) In this letter he suggested that perhaps there was a causal relationship. However, in 1859 it was assumed that space was a complete vacuum and the discovery of electrons and the other subatomic particles was 40 years away. In fact, in 1859 there was still not a clear understanding of electromagnetic radiation (such as light).

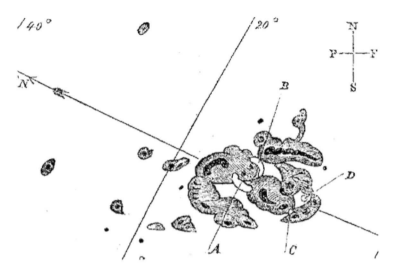

Figure 1.4 Carrington's 1859 drawing of a white light solar flare associated with a sunspot group. (Source: Carrington, 1860)

In 1863 Lord Kelvin,[4] one of the dominant physicists of the nineteenth century, cast strong doubt on the connection between sunspots and geomagnetic activity after calculating that the solar magnetic field could not possibly impact Earth's magnetic field over the tremendous distance between the Sun and Earth. He later expressed these thoughts in his Presidential Address to the Royal Society in 1892: "It seems as if we may also be forced to conclude that the supposed connexion [sic: old British spelling] between magnetic storms and Sun-spots is unreal, and that the seeming agreement between periods has been mere coincidence."

Despite this opinion, several discoveries suggested that perhaps matter from the Sun could travel to Earth. In 1869, J. Norman Lockyer developed the solar spectrograph and observed prominences (large loops of ionized gas) reaching high above the solar atmosphere. The following year Charles Young took the first photographs of a solar prominence. Observations of both solar prominences and total solar eclipses, which showed that the upper atmosphere of the Sun extends away from the solar surface, led scientists to speculate that solar material could be ejected into interplanetary space. Over the next several decades as more observations of activity on the Sun were correlated

---

[4] William Thomson Kelvin (1824–1907), Irish-born physicist and one of the greatest scientists of the nineteenth century. His work included studies in thermodynamics and electricity and magnetism. He was also an entrepreneur and become wealthy with inventions that made the first transatlantic telegraph successful. He proposed the absolute temperature scale, with a degree of measure now called the kelvin (K) in his honor.

with geomagnetic activity, several scientists including Henri Becquerel in 1878 (who later won a Nobel Prize for his discovery of radioactivity) suggested that something (either magnetic fields or matter) occasionally expelled from the Sun impinged upon Earth's upper atmosphere, causing currents to flow and giving rise to geomagnetic activity. As the nineteenth century was drawing to a close, a number of important concepts in fundamental physics that helped in solving the solar–terrestrial puzzle were being developed. These included the realizations that electricity and magnetism are connected, that light is electromagnetic radiation, and the discovery of corpuscular radiation. The discovery of corpuscular radiation in the 1880s by the English physicist William Crookes spurred a number of scientists, among them the Irish mathematician George Francis FitzGerald in 1892 and the English scientist Sir Oliver Lodge in 1900, to suggest that this type of radiation was responsible for geomagnetic activity. At Cambridge University in 1897, J. J. Thomson (1856–1940) showed that corpuscular radiation is made up of subatomic particles called electrons, the discovery of which won him a Nobel Prize. In addition, wireless radio was coming on the scene, and in 1901 the Italian inventor Guglielmo Marconi[5] launched the first transatlantic radio message (which earned him a Nobel Prize). The puzzle of how electromagnetic radiation could propagate around the curved Earth inspired scientists such as the English physicist Oliver Heaviside[6] and the Irish physicist Arthur Kennelly (1861–1939) to theorize that a layer of conducting gas in the upper atmosphere acted to reflect Marconi's radio waves. This theory fit the idea that currents flowing in the upper atmosphere are associated with aurorae. The existence of a conducting layer in the upper atmosphere was soon experimentally verified by the English physicist Edward Appleton[7] (who won a Nobel Prize for his work). The region of the upper atmosphere associated with current flow and the aurorae was named the ionosphere.

[5] Guglielmo Marconi (1874–1937), Italian electrical engineer and inventor who was one of the first to appreciate the potential of radio waves for communication and was the first to send a transoceanic radio message. He developed the first microwave receiver in 1932, which ushered in the development of radar.

[6] Oliver Heaviside (1850–1925), English physicist and electrical engineer who made seminal contributions to the understanding of electric circuits by developing the concepts of inductance, capacitance, and impedance. He rewrote Maxwell's equations of electricity and magnetism in the vector calculus notation used today.

[7] Edward Victor Appleton (1892–1965), atmospheric physicist who played important roles in the development of radar during World War II. He was knighted in 1941 for his contributions in understanding Earth's ionosphere.

Shortly after the discovery of the ionosphere, British geoscientist Sydney Chapman and his student Vincenzo Ferraro published a paper suggesting that streams of particles from the Sun that hit Earth cause geomagnetic storms. Twenty years later, the German astronomer Ludwig Biermann[8] suggested that the Sun continuously emits a gas (now called the solar wind) in order to explain the observation that comet tails always point away from the Sun. Around this time, Swedish physicist Hannes Alfvén[9] proposed the existence of electric and magnetic waves defined by the motion of a magnetized plasma. These waves, now called Alfvén waves, explained how energy and momentum propagate in plasma such as the solar wind (Alfvén won a Nobel Prize in 1972 for this work). Eugene Parker (1927–2022), from the University of Chicago, then developed a theory of the solar wind that explained how the solar atmosphere could continuously expand out into interplanetary space taking with it the solar magnetic field. At Imperial College in London in 1961, James Dungey[10] (1923–2015) combined these ideas by proposing that the solar wind's magnetic field, called the interplanetary magnetic field (IMF), could connect or merge with Earth's magnetosphere and couple solar wind energy and momentum directly into Earth's magnetosphere. Australian solar physicist Ronald Giovanelli (1915–1984) had suggested the concept of magnetic field merging, or reconnection, 10 years earlier in an attempt to explain where the energy came from to power solar flares (like the one observed by Carrington in 1859). These ideas were to be tested over the next 60 years during the space age, when satellites could be sent directly into space to measure what was there. The launch of Sputnik in 1957 ushered in the dawn of the space age and not only brought us understanding of what was in space, but also led to the technological revolution of global communication and Earth observing that we take for granted today. With the advent of satellite and cable TV, Global Positioning System (GPS) navigation, and continental power grid systems, we entered an age in which the turmoil of space began to affect our everyday lives.

---

[8] Ludwig Biermann (1907–1986), German astrophysicist who made important contributions to the understanding of stellar interiors and the understanding of the role of magnetized plasmas in the solar system and galaxy.

[9] Hannes Alfvén (1908–1995), Swedish Nobel-Prize-winning space physicist who founded the subfield of plasma physics called magnetohydrodynamics with his discovery of hydromagnetic waves. These are now called Alfvén waves in his honor.

[10] James Wynne Dungey (1923–2015), British space physicist who developed the idea of how the Sun's magnetic field connects with the Earth's magnetic field through magnetic reconnection to describe magnetospheric dynamics. He is the author's academic grandfather (PhD advisor of my advisor).

1.5 Impacts of Space Weather on Society

In 2020 there were over 3368 operational satellites currently orbiting Earth (this is nearly seven times as many as in 2008 when the first edition of this book was published). Many of these are commercial communications satellites that provide global news TV coverage, telephone connections, and credit card transactions. (Next time you visit a gas station that has "pay-at-the-pump" credit card readers, look at the roof of the gas station. You will probably see a satellite dish that beams your credit card information almost instantly to your bank to verify your credit.) In 2020, investment in commercial space start-ups reached $7 billion dollars, double the investment in 2018, and the commercial space industry is currently a $350 billion industry. Governments operate many other satellites to provide weather images, navigational signals, land-use information, and military surveillance. In the next few years, several commercial companies will be launching thousands of satellites to provide global broadband communication. All of these satellite systems are susceptible to damage and degradation due to the harsh space environment and contribute to orbital debris that increases the chance of operational satellite collisions.

Many other systems, including airline crews and passengers, pipelines, and electric power grids, are also susceptible to space weather effects. Although space weather effects have been observed since the first telegraph lines in the mid nineteenth century, it was not until the 1990s that scientists began studying the problem in earnest. The interest in space weather is primarily due to the rapid growth of the commercial satellite communications industry and the development of continental-sized power and communication grids. With these developments, we have become more susceptible to space weather storms through our reliance on hi-tech information systems and our growing global interconnections. Extreme space weather can damage critical infrastructure, which can lead to cascading failures in other infrastructure systems including power, communication, water, healthcare, and transportation. This book describes space weather causes, effects, and impacts on society. With time, space weather will play a larger role in our everyday lives. Perhaps in the not too distant future, your smartphone weather app will carry a space weather forecast to help you plan your next visit to the orbiting Hilton Hotel or the new resort on the Sea of Tranquility to visit the Apollo 11 landing site on the Moon, or to check the current space weather conditions for the first human colony on Mars.

1.6 Supplements

Measure what is measurable, and make measurable what is not so.

Galileo Galilei

1.6.1 Additional Learning Objectives

After actively reading these supplements, readers will be able to:

- calculate problems of motion using SI units and scientific notation;
- identify the components of physical systems (such as the Earth–Sun system) and identify and organize major concepts that explain the structure and dynamics of each component;
- use active reading techniques to identify key concepts and test their own understanding of new material by asking and answering questions about the material.

1.6.2 Quantitative Understanding

To have a physical understanding of nature, including the environments of the Sun and upper atmosphere of Earth, you must comprehend a number of overarching concepts of physics. Among these are energy and force, which are at the heart of much of physics and help us understand the interconnections and cause-and-effect relationships of the world around us.

In addition to understanding these fundamental concepts of physics, it is also important to learn the language of physics in order to communicate the value of an observed quantity. One could use qualitative terms, such as "fast," "slow," "heavy," "light," etc. as descriptors, but to truly understand an object, you need to know its speed or mass quantitatively. Scientists all over the world use a special set of units called the Système International d'unités (SI), which allows them to communicate easily. You may know SI as MKS (meters–kilogram–second), though the American public still uses "English" units, such as yards, pounds, and seconds. An advantage of the SI (or metric) system is that it is a base ten system (all the units are evenly divisible by 10). In addition, there is a physical relationship between the fundamental units[11] of length, mass, and time. In SI these units are measured

---

[11] Seven units (length, mass, time, electric current, temperature, amount of a substance, and luminous intensity) are generally considered as fundamental. Nearly all other units can be written in terms of these fundamental units and are therefore called "derived units."

in meters, kilograms, and seconds. The relationship between them involves a volume of water and a swinging pendulum (though the latter is only an approximation, and the history of the definition of the meter and second is long and full of twists and turns). The mass of a cubic centimeter of water at standard temperature and pressure (essentially room temperature conditions) is equal to exactly 1 gram. However, the precise conditions made this difficult to reproduce, so early on a chunk of platinum, then rhodium and iridium, was used as a standard. In 2019 the kilogram was defined in terms of a physical constant and the chunk of metal has been relegated to history. One half the period of a swinging pendulum with a length of one meter is almost exactly one second. Today a second is defined by counting the oscillations of a cesium atom and not a swinging pendulum. The development of atomic clocks has made possible a wide range of technologies including satellite navigation. The developers of this technique were awarded the 2005 Nobel Prize in Physics (Roy J. Galuber, John Hall, and Theodor W. Hansch).

1.6.3 SI Units

The base unit of distance is the meter (slightly longer than a yard). Fractions of meters divisible by ten are given prefixes such as deci (1/10), centi (1/100) or milli (1/1000). For distances that are relevant to distances across the surface of Earth, kilometers (km) are often used. This prefix nomenclature is used for all units, so you can have a gram or kilogram or milligram. We often measure time in hours, minutes, and seconds. Therefore, when using time, care must be used in making sure that you use one type of time unit (years, days, hours, minutes, or seconds) and not intermix them.

All of the other fundamental parameters of physics (such as velocity, acceleration, force, and energy) use combinations of the base SI units. For example, velocity (or more properly speed when referring to the magnitude of velocity) is calculated by dividing distance by time ($v = d/t$). Therefore, the units of velocity are the units of distance (m) divided by time (s), or meters per second (m/s). A list of useful SI units and the equivalent base units is given in Appendix 2. Two other fundamental units that are used in this book are for electric charge (coulomb in SI) and temperature (kelvin in SI).

Example: How fast would a coronal mass ejection (CME) need to be traveling if it took 3 days to reach Earth from the Sun?

Answer: Use the velocity formula

$$\text{Velocity} = \frac{\text{distance}}{\text{time}}.$$

The distance between Sun and Earth is 150 000 000 km, therefore

$$\text{Velocity} = \frac{150\,000\,000\,\text{km}}{3\,\text{days}}$$

$$= \frac{150\,000\,000\,\text{km}}{3\,\text{days} \times 24\,\text{h/day}}$$

$$= 2\,000\,000\,\text{km}\,\text{h}^{-1}.$$

Note that days were converted to hours by multiplying the number of days by the number of hours per day.

1.6.4 Scientific Notation

We often measure very large or very small things. For example, the average distance between the Sun and Earth is about 150 000 000 km. Scientists often do two things to make large or small numbers more manageable. The first is to define a new unit. In the case of the distance between Earth and the Sun, we define this distance as 1 astronomical unit (AU) (1 AU = 150 000 000 km). This unit is useful for describing the distance between planets (e.g., Mars is 1.5 AU from the Sun, Jupiter is 5 AU from the Sun, etc.).

Another way of easily handling large or small numbers is to use scientific notation. This is a way to write multiples of 10 using exponents. It is simply a way of keeping track of all the places to the right or left of the decimal point. For example, 1000 can be written as 10^3 and 150 000 000 km can be written as 1.5×10^8 km. For numbers less than 1, the exponent is negative (e.g., $1/1000 = 0.001 = 10^{-3}$). What makes scientific notation useful (besides providing a more compact way to write large or small numbers) is that multiplying and dividing numbers in this form is as simple as adding and subtracting the exponents. To multiply numbers that are written in scientific notation, add all the exponents together. For example, $10^4 \times 10^4 = 10^{(4+4)} = 10^8$. For division, simply subtract the exponents: $10^4 \div 10^6 = 10^{(4-6)} = 10^{-2}$.

Example: The radius of the Sun is 7×10^5 km. The radius of Earth is approximately 7×10^3 km. How many Earths could fit across the face of the Sun?

Answer: The answer can be found by calculating how many Earth radii (r_E) it takes to equal 1 solar radius (R_{solar}).

$$x \; \times \; 1r_E = 1R_{solar}$$

$$x = \frac{1R_{solar}}{1r_E} = \frac{7 \times 10^5 \text{ km}}{7 \times 10^3 \text{ km}} = \frac{7}{7} \times 10^{5-3} = 1 \times 10^2.$$

Therefore 10^2 or 100 Earths can fit across the Sun.

1.6.5 Systems Science

Systems Science is an approach to organizing knowledge and solving problems that breaks down large ideas and problems into different components. Often the organizing principle is based around regions or processes. For example, Earth systems sciences follows the Greek "spheres" naming convention and thinks of the Earth's system as consisting of the different regions or spheres such as the atmosphere, hydrosphere, cryosphere, biosphere, and lithosphere. Scientists use different approaches to understand the system(s), where one approach involves understanding the subsystems or components carefully by themselves and another involves understanding how they interface and interact with other subsystems. The most important systems are global, coupled, and complex: global in the sense of encompassing all the significant systems that interact together; coupled meaning that each subsystem interacts with other subsystems; and complex meaning that the interaction is usually two-way and non-linear. Two-way coupling means that a change in one system influences the other system, which in turn influences the original system. The term non-linear means that the response of one system to perturbations often can grow or damp much faster or drastically compared to the input (e.g., a small change in one parameter can have big impacts on the system). Many complex systems can undergo phase transitions or rapid changes in equilibrium. Ice, for example, is a solid if the temperature is kept below freezing (0 °C), but a small temperature change above freezing can start melting the ice and change the solid into liquid water.

The Earth's climate is an example of a complex, coupled system that requires a global approach for understanding. The coupling and non-linearity often arise due to feedback loops, where a small change in one parameter can lead to rapid changes in multiple systems. The increase in Earth's atmospheric greenhouse gas concentration gives rise to warmer global temperatures because it traps energy that would normally radiate to space without the presence of the gases. Although the total amount of carbon greenhouse gas added to the atmosphere may be small compared with the amount of carbon in the entire Earth system, small changes can warm the Earth's surface, atmosphere, and ocean. These changes can give rise to changes in the amount of moisture in the atmosphere, giving rise to changes in precipitation, severity of storms, and the amount of cloud. Feedback could be negative (e.g., a regulator or thermostat) or positive (e.g., an accelerator), which can drive a system out of balance or equilibrium. One important type of feedback for climate is ice–albedo feedback. As global temperatures rise, the arctic ice cap experiences more days of melting. If the amount of melted ice and snow exceeds the amount of snowfall and ice buildup, the ice thins and the amount of ice decreases. As the amount of ice decreases, more open dark ocean is exposed to sunlight compared with the bright white ice. Albedo is the amount of energy reflected, so an ice-covered ocean has a higher albedo than an open water ocean. Water absorbs heat, contributing to increased melting, which decreases the amount of ice and hence albedo. This feedback can lead to ice-free Arctic Ocean increasing the polar ocean's heat content even more. Arctic ice plays an important role in not only the climate system, but the entire arctic ecosystem with potentially catastrophic consequences to a wide variety of species and native communities.

1.6.6 Conceptual Framework

Systems Science is an approach to understanding physical systems. A learning technique for understanding any system is to develop a **conceptual framework**. This is the way you organize your thoughts and understanding. Making a deliberate choice of how you organize your thoughts, ideas, and understanding can make a big difference in your ability to retain and use information for problem solving. A conceptual framework organizes concepts, content, vocabulary, models, ideas, and techniques into a systematic scheme. This framework helps you understand new knowledge in the context of your existing knowledge and helps you identify connections between different topics. It is a process of learning and primarily consists of creating categories and hierarchies in which all new information can be placed. For example, learning a new area of science (e.g., space weather) can be very hard since it first involves learning a large amount of vocabulary, and new concepts and techniques. To learn or study, many students make long lists of these new words and ideas and then write them out and go over them often. This technique helps them memorize a list of terms as opposed to understanding the material. Attempts to memorize all these new words (perhaps with flashcards or long lists or highlighted terms in the text) are ineffective in the long term as the vocabulary is easily forgotten, given the sheer amount of disconnected information.

Developing a conceptual framework essentially tells a story using the appropriate vocabulary and places the different concepts in an order that shows how they are connected. For example, instead of taking the list of key concepts at the start of each chapter and memorizing them individually, see if you can create a system that organizes and connects them. Is there an analogy that you already know that can be used to understand the new concept? For example, one framework that is used in this text is thinking about "regions," "processes," and "events." Regions discussed in this chapter include the Sun, the solar wind, the magnetosphere, and the ionosphere. If you look ahead or at the contents list you will notice that many of the chapters are devoted to these regions. Notice that the "connection" of these regions is through processes briefly introduced in Section 1.4.4 such as energy and magnetic field merging or reconnection. These processes are discussed throughout the textbook.

Finally, we often focus on and study rapid changes or dramatic "events": hurricanes, economic recessions, heart attacks, automobile accidents, the onset of war, elections, etc. In this chapter we introduced space weather events such as solar flares, CMEs, geomagnetic storms and substorms. To describe and understand these events, we need to place them within a region and understand how processes in that region and other regions that are connected to it give rise to the conditions for the event.

Developing conceptual frameworks is a powerful learning strategy and is used by economists, medical doctors, art historians, musicians, psychologists, athletes, and in fact every area of human endeavor. Unfortunately, most students never formally learn this approach. Several supplements at the end of future chapters describe this and other related learning techniques that will help you understand space weather and can be used for all of your classes and current and future endeavors.

1.6.7 Active Reading

How do you read textbooks and other technical communication? For some students, if the assigned reading is read at all, it is read in the same fashion as any other genre of written communication – from the beginning to the end in an attempt to memorize the important information, definitions, vocabulary, characters, plot components, etc. as you go along. Technical reading should be approached differently to help identify, understand, synthesize, and place the material into your cognitive or conceptual framework. **Active reading** is an approach that engages the reader with the text to explicitly identify the key information and understand where it fits into your previous understanding. The first part of active reading is to define the purpose or goal of the reading. Are you trying to learn new material, identify the main concept, learn a new technique, develop new vocabulary, or understand the relationship between a new concept with an earlier concept? With that goal in mind, the second part of active reading is to have a pen or pencil or text editing capability on a pdf to write directly in the margins or in your notes.

Regardless of the way you are writing notes – physically in the margins with a pencil or digitally annotating a pdf – the technique of active reading is the same. After identifying your goal or purpose, find and read the relevant passages. For example, if reading a chapter for the first time, your purpose may be to get a broad overview of the topic. The chapter title usually is a good indicator of the main topic of the chapter. Chapter subheadings are good clues on the hierarchy or way that the broad topic can be organized or subdivided into smaller components. After each subsection you should write in the margins a one-sentence summary of the main point, and one question that you have about the reading. At the end of the chapter write another sentence or two as a summary of the topic connecting some of the subtopics into the overview. Writing questions about where the new material fits in with previous chapters or content, and comparing or contrasting subtopics discussed, are good active learning strategies. Can you identify why the topic of the chapter was divided into these subtopics? Write the questions and answers in your notes. The development of questions and answering them is a major component of active reading; it helps place the information into your long-term memory and make connections across different ideas.

When you review material, the purpose should be to understand the main concepts and techniques. Try to summarize the main points of the text in a few sentences. Then develop questions that would test your understanding of these concepts or techniques. For example, one of the main points of Chapter 1 was to introduce and connect sunspots, geomagnetic storms, and aurorae. Can you tell a story that connects these three phenomena together? After developing a set of questions, you may find that they are the same ones on your quizzes or exams.

1.7 Problems

1.1 Have you used a satellite recently? Think about how you have used information from a satellite or gained access to information using a satellite and write a short paragraph describing this use.

1.2 What is the name of the boundary that separates Earth's magnetosphere from the solar wind? How would you tell when you crossed this boundary?

1.3 If a communication company launched 1000 satellites at a cost of $10 million each (to build, launch, and operate), how much revenue would they need to recoup their costs over a three-year period? At $50 per month, how many customers do they need? Is this a realistic business plan? (Verizon is the largest cellular company in the USA, with over 145 million customers in 2019, and several satellite communication companies, such as Elon Musk's Starlink and Jeff Bezos' Kuiper Systems, are attempting to develop global space-based communication systems with potential for hundreds of millions of customers.)

1.4 What is the distance from the Sun to Earth in terms of solar radii? Earth radii?

1.5 How long (in days) does it take a parcel of solar wind traveling at 400 km s^{-1} to reach each planet? (Mercury = 0.4 AU, Venus = 0.7 AU, Earth = 1.0 AU, Mars = 1.5 AU, Jupiter = 5 AU, Saturn = 10 AU, Uranus = 20 AU, Neptune = 30 AU, Pluto = 40 AU [yes, Pluto is no longer a planet, but …])

1.6 How does the development of new observational instruments contribute to our understanding of our natural world?

1.7 What was the state of long-distance communication technology in 1859? Could there have been any space weather technology impacts to communication technology due to the 1859 Carrington geomagnetic storm? Research this question and summarize your findings in a few sentences. Provide citations to sources.

1.8 Write a sentence or two summarizing one of the sections of Chapter 1 and one question that would probe your understanding of the main point of the section. Then answer your question.

1.9 What are some of the subsystems that are part of the commercial aviation industry and how are they inter-related? Explore how changes in one component or subsystem influence another and how that change could in turn influence the original subsystem.

1.10 Either do research and describe an example of a cascading failure or write a short story imagining how the failure of a simple electronic circuit on a communication satellite could cause a disaster on Earth.

Chapter 2
The Variable Sun

the spots do not remain stationary upon the body of the sun, but appear to move
in relation to it with regular motions
  (Galileo Galilei in a letter to Mark Welser, 1613, quoted in Drake, 1957)

Galileo's "Sunspot Letters" were one of the first written scientific
discussions of sunspots.

2.1 Key Concepts

- electromagnetic radiation
- heat transfer
- solar atmosphere
- solar cycle
- Standard Solar Model

2.2 Learning Objectives

> After actively reading this chapter, readers will be able to:
>
> - define the physical processes and characteristics of regions of the
>   Sun;
> - describe the different forms of energy that are associated with
>   solar regions, features, and events, and organize them based on
>   their spatial and temporal scales;
> - develop a concept map of the Sun.

2.3 Introduction

Since the dawn of human life, the Sun has elicited worship, inspiration,
and study, and tremendous mysteries about its dynamics still occupy the
attention of solar astronomers and space physicists. This ignorance has
profound implications. We now rely on space for global communication, navigation, and Earth observing, and solar dynamics causes degradation and failure of satellites, space instruments, and ground-systems
such as power grids. Understanding solar dynamics is a key part of
understanding space weather. This chapter describes what we know

about the Sun and how we know it. Much of our knowledge comes from observations of the Sun, and much of it comes from applying the laws of physics (such as thermodynamics and nuclear and atomic physics) within quantitative models to make predictions of observable quantities. This is how we know what goes on inside the core of the Sun without going there or observing it directly. This combination of observation and physics has allowed us to know more about our natural surroundings than at any other time in human history. What we are finding out is that we know very little of how things work. But what we do know (and how we have come to know it) is truly amazing.

The Sun is one of an estimated 100 billion stars in the Milky Way galaxy. We know this by extrapolating surveys of stars in this galaxy and others. From studying the properties of nearby stars and the age of meteorites, we have learned that the Sun is a typical star about 4.5 billion years old. By observing star formation regions within our galaxy, we know that the Sun was formed out of a giant cloud of gas and dust called the solar nebula. An area within this nebula contained slightly more material than its surroundings and therefore gathered material to itself because of its gravitational pull (a self-sustaining process called accretion). As the proto-Sun ("proto" being Greek for "first") got bigger and bigger, its increasing mass strengthened its gravitational attraction, pulling more material to it. As the gas and dust were gathered into the proto-Sun, gravity forced it closer and closer together, increasing its density (the amount of material in a given volume). The pressure inside the proto-Sun therefore increased since the pressure of a gas is dependent on its density, which we know from the ideal gas law (expressed as $P = nkT$, where P is pressure, n is number density, k is Boltzmann's constant,[1] and T is temperature). As more material was accreted onto the proto-Sun, its density and temperature continued to increase until they reached a critical temperature at which thermonuclear fusion can occur (about 15 million K). At this point the Sun was born. A star becomes stable when there is a balance between the force of gravity pulling material into the star and the force due to the pressure of the gas pushing out. This is called hydrostatic equilibrium (see Chapter 5 for further discussion of this concept). The Sun's energy comes from thermonuclear reactions in the core that fuse protons together to form helium nuclei. In the process, energy is liberated;

---

[1] The Boltzmann constant was named in honor of Ludwig Boltzmann (1844–1906), an Austrian theoretical physicist who made significant contributions to thermodynamics and electricity and magnetism and is credited with developing the foundation of statistical physics through his insights into the kinetic theory of gases. His formula describing the relationship between the mean total energy of a molecule and its temperature includes a constant k, now known as the Boltzmann constant (equal to 1.38066×10^{-23} joules per kelvin).

some of this energy eventually makes its way to the surface and propagates out into space as **electromagnetic radiation**.

Because of the high temperatures on and within the Sun, the gas is ionized. An atom or molecule becomes ionized when an electron is knocked out, giving the atom a net positive electric charge. The charged particles feel the force of electric and magnetic fields. In addition, moving charged particles can create an electrical current, which in turn gives rise to a magnetic field. Therefore, due to the motion of ionized gas in the Sun, a strong magnetic field is generated. Changes in this highly variable solar magnetic field cause changes in the amount of energy released from the Sun's surface. The Sun is a dynamic and variable star because of these moving charged particles and the magnetic field they create. The structure and processes of the Sun that give rise to solar variability are examined in this chapter.

Before discussing the atmosphere of the Sun, a few important physical concepts to help explain the Sun's structure are introduced.

2.4 Temperature and Heat

The temperature of an object, gas, or liquid describes the thermal motion of the atoms and molecules that make up the object. Atoms or molecules in hot things have large thermal velocities, and atoms or molecules in cold things have small thermal velocities. A low-temperature gas or liquid has slowly moving molecules; a high-temperature gas or liquid has quickly moving molecules. Put a tea bag (or a drop of food coloring) in hot water and then another in cold water. Does the tea (or food coloring) spread more quickly in hot water or cold water? What does this say about the motion of the individual water molecules in the cup?

When objects with different temperatures are placed in contact with each other, the temperature of both objects changes. The hotter object cools while the colder object warms. The temperatures change until they are in equilibrium or equal. We call the transfer of energy from a hot object to a cold object heat. Heat is measured in units of energy (joules in SI units), and the transfer of heat from one place to another drives weather on Earth and is a fundamental aspect of space weather.

2.5 Radiation and Convection

There are three forms of **heat transfer**: conduction, convection, and radiation. Conduction is the transfer of heat in the absence of fluid motions. This occurs when two solid objects are brought into contact (e.g., an electric stove top and a frying pan). The hot stove top's heat flows from the electric coils into the metal frying pan, heating the frying pan.

Convection is the transfer of heat by fluid motions. Fluid is a generic name given to everything that flows easily, and fluids can include gases. An example of convection is how heat is transferred through a pot of boiling water. The hotter water near the bottom in contact with the hot pot rises because it is hotter and hence less dense; the cooler water sinks because it is more dense. If you watch a pot of water boil, you will see convection cells set up with the hot water rising at the center of the pot and sinking at the sides.

Radiation is the transfer of heat through electromagnetic radiation. The Sun warms Earth's surface when sunlight is absorbed by Earth. The temperature of Earth depends on the amount of sunlight absorbed by Earth and its atmosphere.

These last two processes (convection and radiation) are responsible for the transfer of heat from the core of the Sun into outer space. The heat from the Sun's core is transferred towards its surface in the form of electromagnetic radiation. At a certain point inside the Sun, fluid flow can effectively start transferring that heat via convection. The gas is heated from below and rises, reaches the surface, emits radiation (and hence heat) out into space and cools and sinks.

2.6 Solar Structure

We can make direct observations of the solar surface and atmosphere. The primary means of study of these regions is analysis of absorption lines in the solar spectrum. By studying these lines we know the composition of the Sun to very high accuracy. Table 2.1 lists the five most common elements in the Sun and their relative abundance. Hydrogen is by far the most common, followed by helium. The

Table 2.1 *The five most common elements in the Sun*

Element	Symbol	Relative abundance
hydrogen	H	92.1%
helium	He	7.8%
oxygen	O	0.061%
carbon	C	0.030%
nitrogen	N	0.0084%

Standard Solar Model suggests that these abundances are representative throughout the Sun except in the core where thermonuclear reactions continuously change the composition. The Sun contains all of the natural elements found on Earth and in the periodic table. In fact, essentially all of the elements beyond hydrogen and helium on Earth (and in our bodies) were made inside a now-dead star, whose remains made up the original solar nebula. Literally, we are made of star-stuff.

The Sun contains 1.9×10^{30} kg of material – over 99% of the total mass in the solar system – or about 300 000 Earth masses. The regions of the Sun are illustrated in Figure 2.1. Table 2.2 shows some of the Sun's physical characteristics.

Table 2.2 *Solar properties*

radius	696 000 km
mass	1.9×10^{30} kg
average density	1410 kg m^{-3}
distance from Earth	150 000 000 km (or 1 astronomical unit)
surface temperature	5800 K
luminosity	3.86×10^{26} W

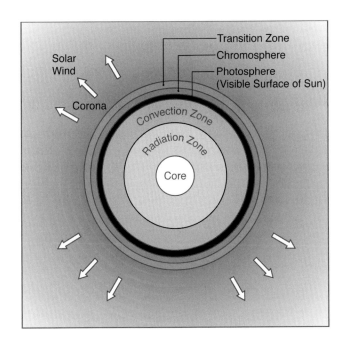

Figure 2.1 The main regions of the Sun. The regions inside the Sun are defined by how energy is transferred from the core to the surface. The regions of the Sun's atmosphere are defined by their density and temperature.

2.6.1 Interior

The Sun's interior is divided into three main regions: the convection zone, the radiation zone, and the core. The next sections describe these regions.

Convection and Radiation Zones

Below the photosphere, the visible surface of the Sun, extending down about 200 000 km, is the convection zone. This region undergoes convective motion (hot gas rising and cooler gas sinking) similar to a pot of boiling water. Convection is the process that transports heat through bulk fluid motion. Energy from the outer layer of the Sun is transported by convection to the solar surface where it can radiate out into space. Below the convection region is the radiation zone, where energy is transported primarily by electromagnetic photons. The convection zone begins where the flux of radiation energy is so high that the energy cannot easily make its way through the gas, hence convection begins in order to transport the energy through bulk motions of the solar material.

How do we know what goes on inside the Sun? As previously mentioned, we have developed a solar model (the Standard Solar Model) based on our best understanding of the physics that takes place there. The model is compared with observations that we can make – including direct observations of the solar surface – which allows measuring in-and-out surface motions. These radial motions or oscillations are observed through periodic changes in the Doppler[2] shifts of spectral lines of the gases in the photosphere and chromosphere. The observed spectral frequency (or color for visible light) is shifted toward the blue for gas that is moving towards the observer and toward the red for gas moving away (see supplements in Section 2.8 for details). Solar surface motions are due to sound waves from the interior of the Sun. Different regions of the Sun refract (bend) or reflect these sound waves and confine the motion of the waves to specific regions. Observers can then deduce the temperature and density structure of the interior from the behavior (frequency) of these oscillations. This is similar to how we can use seismic waves to deduce the structure of Earth's interior. Because of these similarities the field of using sound waves to study the interior of the Sun is called helioseismology.

---

[2] Christian Johann Doppler (1803–1853), Austrian physicist who discovered that the observed frequency of a sound wave depends on the relative velocity of the source and observer. This is called the Doppler effect, and it applies to electromagnetic radiation (such as light) waves as well, which allows the remote sensing of the solar surface's radial velocity.

The Core

Thermonuclear reactions occur in the Sun's core, which is about
200 000 km across. These reactions power the Sun and release tremendous
amounts of energy every second. We can measure the Sun's total energy
output by placing a device – such as a solar cell – above Earth's atmosphere
perpendicular to the Sun's rays and determining how much energy is
intercepted per square meter every second. This quantity of energy, called
the solar constant, is about 1400 W m^{-2}. An average US residential home
uses 1200 W,[3] so the energy intercepted by less than one square meter could
provide the electrical energy of an average household.

Now that we know how much solar energy passes through a square
meter above Earth's atmosphere per second, we can calculate the total
amount of energy emitted by the Sun each second. This quantity, called
luminosity, has units of power. Assuming that the Sun radiates energy
uniformly in all directions, we can imagine that the amount of energy
intercepted at one square meter above Earth's atmosphere is the same
amount of energy intercepted at any square meter of an imaginary
sphere that has at its center the Sun and has a radius equal to the distance
from the Sun to Earth. Now all we have to do is add up all the square
meters that cover the surface of that sphere. The surface area of a sphere
is $4\pi r^2$. Therefore the total amount of energy emitted by the Sun per
second is the solar constant times the surface area of a sphere whose
radius is the distance between the Sun and Earth (1400 W m^{-2} × 4π
(1 AU)^2), approximately 4×10^{26} W. Although the luminosity of the
Sun is typical of the other stars you see in the sky, the Sun appears much
brighter (and bigger) because it is much closer to us than even our
nearest stellar neighbor. To put its luminosity in perspective, the Sun
emits the same amount of energy in one second that Earth would
produce at the current rate in over 900 000 years. Put another way, the
Sun emits the equivalent of 100 billion one-megaton nuclear bombs
every second. Needless to say, nothing on Earth compares with the vast
amounts of energy produced each second by the Sun.

A natural question that had puzzled scientists until the twentieth
century is, where does the Sun get this energy? One of the biggest
contributions to our understanding of the Sun's energy source was
Einstein's[4] famous equation, $E = mc^2$, which tells us that energy equals

---

[3] Residential energy use is metered and billed in kilowatt-hours (kW h) or the number of
kilowatts used each hour. In 2017, a typical US home used 10 399 kW h or an average of
1200 W. Of course not all the energy from the Sun makes it through Earth's atmosphere.
Some of it is absorbed or reflected by the atmosphere and clouds, so on average only
30–50% of the energy makes it to Earth's surface.

[4] Albert Einstein (1879–1955), German-born American Nobel-Prize-winning physicist
whose work revolutionized our understanding of energy and matter.

mass times the speed of light squared or, physically, that mass and energy are intimately related. The speed of light is itself a very large number, $3 \times 10^8 \, \text{m s}^{-1}$, and this number squared is huge. Therefore, even a small amount of mass is equivalent to a large amount of energy. Einstein's simple formula led to the development of the concept of nuclear fusion, now known to be the only energy-generating mechanism capable of producing the enormous amounts of energy emitted by the Sun. In this process, light nuclei are fused into heavier nuclei. High temperatures in the Sun's core strip electrons from nuclei of atoms (mostly hydrogen) so that there are essentially protons (or the nuclei of hydrogen) whipping around the core. Occasionally two protons collide, which starts a process called a proton–proton chain that eventually leads to one helium nucleus. Helium contains two protons and two neutrons and therefore has an atomic mass of 4. It is made by fusing together four protons. If you compare the mass of four protons with the mass of a helium nucleus, you find that there is a discrepancy. The helium nucleus is slightly lighter. Where did the missing matter go? It went into energy, according to Einstein's formula. The difference in mass is about 0.0477×10^{-27} kg – not a lot – but when converted to energy it is equivalent to 4.3×10^{-12} J. Therefore, fusing together 1 kg of hydrogen (and converting the tiny fraction of its mass into energy), can generate 6.4×10^{14} J (or the amount of energy Earth would generate at current levels for the next 1600 years – now you can appreciate the promise of nuclear fusion for solving our energy needs). To generate enough energy to account for the Sun's current luminosity, we must convert 600 million tons of hydrogen into helium each second. This is a lot of mass, but very little compared with the total mass of the Sun. The energy from nuclear fusion is emitted in the form of gamma rays – the highest-energy form of electromagnetic radiation. As gamma-ray photons make their way through the Sun – colliding and being absorbed and re-emitted by the matter in the Sun – they lose energy. Eventually the energy leaves the photosphere, mostly in the form of visible light. A small amount of the energy is carried off by neutrinos as a by-product of the fusion process. These subatomic particles (the name "neutrino" derives from the Italian for "little neutral one") move off at essentially the speed of light and effectively escape the Sun without any interactions. They are so weakly interacting that they pass through you and Earth continuously (in fact they can pass through several light-years of lead). However, we have developed sensitive instruments that can detect a very small fraction of the neutrinos intercepting Earth. These measurements have helped confirm our understanding of the processes in the core of the Sun.

2.6.2 Solar Atmosphere

Like the interior of the Sun, the **solar atmosphere** is divided into three layers – the photosphere, the chromosphere, and the corona. As in the Earth's atmosphere, each layer is characterized by different optical properties, temperatures, and temperature profiles with height above the surface. The next sections discuss the photosphere and chromosphere, and the following chapter discusses the corona.

The Photosphere

The visible surface of the Sun – the photosphere ("photo" from the Greek for "light") – is opaque to visible light, and hence we see a sharp edge. However, since the Sun is made of gas, it does not have a solid surface. Galileo first systematically observed the visible surface of the Sun in about 1610. Figure 1.3 is a drawing of Sun's surface made by Galileo showing that the Sun contains spots (whose number waxes and wanes over an 11-year **solar cycle**). These sunspots are now known to be regions of strong magnetic field. Sunspots appear darker than the surrounding solar surface because they are slightly cooler (typically 4500 K compared with the 5800 K of the surrounding photosphere). They are usually about 10 000 km (1–3 Earth radii) across and can last for weeks. As mentioned in Section 1.4.3, sunspots were used by Galileo to estimate the Sun's rotation rate. The Sun takes about a month to make a complete rotation. Richard Carrington used sunspots, 250 years after Galileo first discovered solar rotation, to determine that the Sun rotates differentially (i.e., moves at different speeds at different solar latitudes). The Sun's equator rotates faster than the Sun's poles: the equatorial rotation rate is about 25 days, while the poles rotate every 36 days. The average rotation rate that has greatest space weather impact on the Earth is about 27 days.

High-resolution images of the photosphere (Figure 2.2) show that the Sun's visible surface is highly mottled, with dark and light regions called granules. Granulation is direct evidence of solar convection. Bright areas show a Doppler blue shift indicating upward motion. Dark areas show a red shift indicating motion down into the solar interior. The bright regions are about 1000 km across (or about the size of the state of Texas in the USA). Bright spots come and go with a lifetime of about 5 to 10 minutes. As with sunspots, the dark regions are typically about 500 K cooler than the bright spots.

In addition to this fine-scale granulation there are observations of supergranulation with scale-sizes of tens of thousands of kilometers (or several Earth radii). These super granules are thought to be the signature of larger-scale convection in the outer convective zone.

Figure 2.2 High–resolution image of the photosphere showing a sunspot. The smaller features, called solar granulation, provide evidence for convection. (Source: Friedrich Woeger, KIS, and Chris Berst and Mark Komsa, NSO/AURA/NSF)

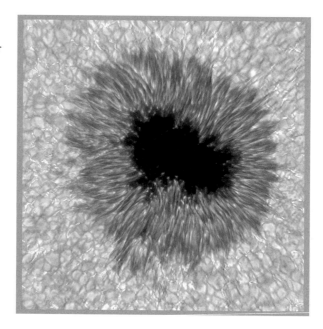

The Chromosphere

The region just above the photosphere is called the chromosphere ("chromo" from the Greek for "color"). The chromosphere, which is about 1500 km thick, is characterized by a temperature higher than that of the photosphere (about 10 000 K compared with the 5800 K temperature of the photosphere) and appears reddish in color during total solar eclipses, giving the region its name. The element helium ("helios," Greek for "Sun") was discovered in the chromosphere before it was discovered on Earth. Density of the plasma (and hence the amount of light emitted) drops rapidly with height in the chromosphere. Therefore the chromosphere is not visible against the bright background of the photosphere. However, scientists have known about the chromosphere for a long time by observing total solar eclipses. When the Moon blocks the bright photosphere, the chromosphere becomes visible with its distinctive reddish hue due to the H-alpha (Hα) emission line of hydrogen. Hα, which refers to a specific electronic transition within hydrogen, is associated with a unique wavelength of visible light. The chromosphere can be very dynamic with hot jets of gas (spicules) extending high above the surface. These can extend thousands of kilometers above the solar surface and are made of material that has been expelled from the surface at velocities of about 20–100 km s^{-1}. The chromospheric temperature is inversely proportional to density, so it increases rapidly with height.

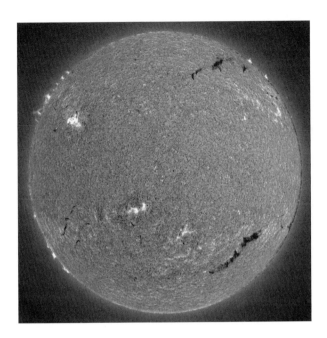

Figure 2.3 The Sun observed in the light of hydrogen-alpha. (Courtesy of Big Bear Solar Observatory)

When the Sun is viewed through a special filter that allows only Hα light through, a wealth of features become apparent (Figure 2.3). Chromospheric networks are web-like patterns most easily seen as bright emissions in Hα. These networks outline the supergranulation cells mentioned above. Spicules and solar prominences tend to be found at the edges of these cells. Very narrow dark filaments and bright filaments (called plages – French for "beach") also are visible in Hα. Filaments observed above the limb of the Sun, called prominences, are often characterized by their loop-like appearance. The loops map out the solar magnetic field, and the magnetic field is what constrains the motion of the bright gas.

Above the chromosphere the temperature increases dramatically with height in a relatively thin transition region. Above this is the outer atmosphere of the Sun, called the solar corona, which is visible during solar eclipses and from special telescopes called coronagraphs. The corona (Latin for "crown") expands out into space supersonically. The solar gas that escapes into interplanetary space is called the solar wind. This region will be discussed in more detail in the next chapter.

2.7 Dynamics and Processes

2.7.1 Solar Magnetism

The magnetic field strength of various regions of the Sun can be determined spectroscopically using the Zeeman effect, discovered by

Pieter Zeeman[5] in 1896. In the presence of a magnetic field, a gas's spectral lines will be split into two or more components. The actual frequency of the spectral lines depends on the strength of the magnetic field. This discovery won Zeeman and his former teacher (Lorentz)[6] the Nobel Prize in Physics in 1902 and allowed direct measurement of the solar surface magnetic field for the first time. Using this technique, we know that the magnetic field inside a sunspot is about 1000 times stronger than in the surrounding normal solar surface. In addition, the fields are directed either into or out of the solar surface and not randomly oriented. Magnetic fields have a direction and magnitude. For example, a magnet has a north and south pole, with the north pole defined as the side that has magnetic field moving away from the magnet, and the south pole where the field is directed towards the pole. It is thought that the strong fields associated with sunspots interfere with the normal convection of the Sun and hence the gas can cool and appear darker than its surroundings. Each sunspot generally has a single polarity (or magnetic field direction), and sunspots usually appear in pairs of opposite polarity. The north polarity (or outwardly directed magnetic field) makes a loop about the solar surface and then dives into the south polarity (inward directed) sunspot. Although sunspots often have an irregular appearance, the magnetic field orientations exhibit a great deal of order. All the sunspot pairs in a given hemisphere of the Sun have the same magnetic configuration. Specifically, when the leading spot (measured in the direction of solar rotation) has a north polarity, then all leading spots in the hemisphere will have a north polarity. What is more, all the sunspot pairs in the opposite hemisphere will have the opposite orientation (south leading). This ordering of the solar field is due to the differential rotation of the Sun. Recall that the Sun rotates faster at its equator than at the poles. As seen in Figure 2.4, this differential rotation causes a twisting of the overall solar magnetic field.

2.7.2 Solar Active Regions

Many pairs of sunspots are associated with explosive releases of energy from the photosphere. These areas of activity are simply called active regions. Although the exact mechanisms that cause the explosive

---

[5] Pieter Zeeman (1865–1943), Dutch Nobel-Prize-winning physicist who discovered that spectral lines split in the presence of a strong magnetic field. This is called the Zeeman effect and allows the remote sensing of the strength of the Sun's magnetic field.

[6] Hendrik Antoon Lorentz (1853–1928), Dutch Nobel-Prize-winning physicist who helped develop the theory of electromagnetism and predicted the Zeeman effect.

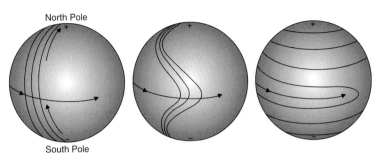

North Pole

South Pole

Figure 2.4 The Sun rotates faster at the equator than at the poles. This is possible because the Sun is a ball of gas. The three panels show the effect of differential rotation on the Sun's magnetic field. (Source: NASA TRACE mission)

release of energy from the Sun's surface are not known, we do know that they are related to the rapid conversion of magnetic energy into particle kinetic energy. This conversion takes place in regions of strong magnetic fields, and the twisting of the surface magnetic field often leads to rapid energy release. Prominence eruptions are an example of this energy release. Another example, solar flares, are much more energetic than prominences. Flares release tremendous amounts of energy in a few minutes and can reach temperatures of 100 million K (much hotter than even the core of the Sun). This energy is equivalent to hundreds of millions of megaton hydrogen bombs exploding at the same time. The energy of these flares is so intense that the charged particles that make up the solar atmosphere are blasted out into space – some at nearly the speed of light. In addition, the heated gas glows at essentially all wavelengths including X-rays. These energetic particles and electromagnetic radiation are ejected into interplanetary space and can often impact Earth's space environment – one of the causes of space weather.

2.7.3 Solar Cycle

The number of sunspots on the Sun changes over an 11-year cycle. Over this solar cycle, the number of sunspots waxes and wanes. Since sunspots are associated with solar activity (i.e., flares and other rapid releases of energy that can heat localized regions of the atmosphere of the Sun to many millions of kelvin), the solar cycle also describes the level of activity and variability of the Sun. Plate 3 shows an extreme UV photograph of the Sun taken by the European PROBA2 satellite over an 11-year solar cycle. Each snapshot of the Sun shows the structure of the Sun's atmosphere in UV at a given day, one snapshot per year. Note that the UV radiation increases and decreases over this solar cycle – with the amount of time from the most active or energetic interval (called solar maximum) to the time of least activity (solar minimum) being approximately 5–6 years. UV is very energetic electromagnetic radiation

Figure 2.5 The number of sunspots observed on the Sun since the 1700s. Note the approximately 11-year periodicity of the occurrence of sunspots, which is called the sunspot cycle. (Source: WDC-SILSO, Royal Observatory of Belgium, Brussels)

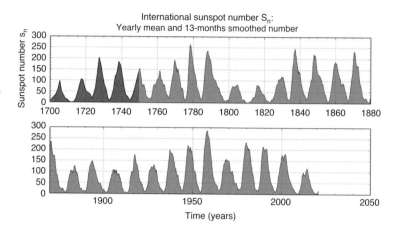

created when a gas is heated to tens of thousands of kelvin. Notice that during solar maximum the Sun's atmosphere has many places of bright UV emission, whereas at solar minimum the atmosphere of the Sun is dim in UV, indicating that very little heating of the upper atmosphere of the Sun is taking place. Also note that the bright UV emission is not coming from the surface of the Sun, but from the atmosphere above the photosphere.

This cycle has persisted for centuries. Figure 2.5 shows the number of sunspots observed on the Sun since the 1700s, when sunspots were observed on a daily basis from a number of different solar observatories. The number of sunspots on the Sun changes continuously – sunspots typically have lifetimes from 1 to 100 days – but the total number of sunspots changes with this quasi-regular 11-year cycle.

Recall that sunspots are regions of intense magnetic fields, and the variation of their number indicates that the Sun's magnetic field is also changing. In fact the structure and orientation of the Sun's magnetic field changes over a 22-year cycle, with the polarity of the Sun's field reversing over this time interval. The Sun's magnetic field is generated and modulated by the Sun's dynamo, which is powered by the differential rotation of the Sun and the solar convection processes. A dynamo is the mechanism that generates magnetic field. During solar minimum the Sun's field is relatively simple and well ordered, resembling a dipole magnetic field, with the magnetic field coming out of one hemisphere and going in the other. Over the next 5–6 years as the Sun approaches solar maximum, the nice dipole configuration slowly disappears and the Sun becomes magnetically disorganized and highly structured. After solar maximum, over the next 5–6 years the magnetic field again becomes more structured and more dipolar. Early in this transition the tilt of the weak dipole field can be very large with respect to the spin axis of the Sun, but as solar minimum approaches the dipole axis orientation becomes more and more aligned with the spin axis. When the dipole field is reformed, it has the opposite polarity than the previous one. This change in polarity is what defines the 22-year magnetic cycle of the Sun, sometimes called the double solar cycle or the Hale[7] cycle. The polarity of sunspot pairs also follows this cycle. For the first 11 years of the magnetic cycle, the leading spots of a particular hemisphere always have the same polarity, which are opposite to the polarity of the leading spot in the other hemisphere. The sunspot polarities then reverse for the next 11 years.

Since the amount of solar activity follows this sunspot or solar cycle, one would expect that the number of solar disturbances that impact Earth would also follow this cycle. This is exactly what happens. Therefore space weather has "seasons," with solar maximum indicating a strong likelihood of severe space weather, and solar minimum predominantly quiet space weather. The next chapter describes the outer atmosphere of the Sun – the solar corona and solar wind. Earth resides in the outer atmosphere of the Sun and so our space environment is intimately connected to the structure and dynamics of the Sun.

2.8 Supplements

2.8.1 Additional Learning Objectives

After actively reading these supplements, readers will be able to:

- list the different frequency bands of the electromagnetic spectrum, calculate wavelength and energy from frequency, predict the observed change of frequency of an object due to its relative motion toward or away from the observer, and describe how we can derive an object's temperature by observing the amount of electromagnetic emission as a function of frequency;
- make a concept map of electromagnetic radiation and other solar regions, processes, and events and use these concept maps to probe their understanding of new concepts and how they fit within their own previous knowledge.

---

[7] George Ellery Hale (1868–1938), American solar astronomer instrumental in founding a number of astronomical observatories.

2.8.2 Electromagnetic Spectrum and Radiation

Frequency and Wavelength

Many things in nature, such as a pendulum, an ocean buoy, the walls of our hearts, and a piano string, move back and forth or up and down in a regular manner (are said to oscillate). Some oscillations perturb their surroundings and create waves, and the perturbations or waves carry energy from their source to their surroundings. A piano string, for example, creates sound waves that perturb the surrounding air's density and push it back and forth as the string oscillates. Our ears are sensitive to these air density fluctuations, and we can hear the sound waves.

A characteristic of these waves is the time between oscillations (a cycle), or in the case of a sound wave generated by a piano string, the time it takes the string to move back and forth or the time it takes one density perturbation cycle to pass us. We call this time the period (T). A closely related parameter called frequency (f, sometimes written as v) is the number of events (or in the case of the piano wire the number of back-and-forth oscillations or cycles) in a given time interval. Period and frequency are related by $f = (1/T)$. The SI unit of frequency, equivalent to one cycle per second, is the hertz[8] (Hz).

One way to visualize a wave graphically is to record the displacement (or amount of motion) of the string as a function of time, or, in other words, plot the string's position (along the y-axis) as a function of time (along the x-axis). Figure 2.6a shows the position of a point at the middle of the piano string after it has been struck as a function of time. Note that the string moves back and forth about its resting position. The amount of displacement (a measure of the size of the wave) is called the amplitude. The time between crossing points, crests, or troughs on the curve is the amount of time to complete one cycle or the period.

Now let's examine a sound wave created by the vibration or oscillation of a piano string. Instead of plotting the position of the string as a function of time, we could record the density of the air at a given point next to the string as a function of time. This plot would look just like the position of the string as a function of time. This is because when the string is moving towards our microphone or ear, the air is compressed and its density goes up. When the string moves away from our microphone, the air is rarefied and its density goes down. The timing between the air density changes is exactly the same as the string's back-and-forth motion that created them. Therefore the traveling sound wave also can

---

[8] Heinrich Rudolf Hertz (1857–1894), German physicist whose study of electromagnetic radiation led to the discovery of radio waves. The SI unit of frequency is named in his honor.

be described by its period or frequency and amplitude. In fact, all periodic waves can be described in exactly the same way.

Recall that waves are disturbances that move away from their source. A water wave moves away from your hand as you splash in a pool; a sound wave moves away from a piano string. How fast do these waves move? The answer depends on the density and temperature of the material the wave propagates through. A wave will move at different speeds through different materials (sound will travel at different speeds through air and water, for example) and its speed also depends on the temperature of the particular material (i.e., sound waves move faster through warm water than through cold water).

There is a relationship between a wave's frequency, wavelength, and velocity. The velocity is equal to the wavelength times the frequency

$$V = \lambda f.$$

Wavelength describes the distance between the crests and troughs of the wave. So another way to visualize a wave similar to Figure 2.6a is to record the amplitude as a function of distance (instead of time) as done in Figure 2.6b. It has the same sinusoidal shape with the crests and troughs at any given moment in time representing the spatial structure of the wave. The distance between the crossing points, crests, or troughs is the wavelength.

For light, the velocity is written c (as in Einstein's famous equation), and the energy of the electromagnetic radiation depends on the frequency (or wavelength). High-frequency EM radiation has more energy than low-frequency radiation. This linear relationship can be written

$$E = hf,$$

where h is Planck's constant[9] and E is energy. So, for visible light, red is lower frequency than blue light and therefore red light carries less energy than blue light. Figure 2.7 shows the frequency and wavelength of the entire EM spectrum.

Doppler Effect
The Doppler effect is the shift in frequency of a wave due to the relative motion of the sound emitter and observer. For example, as a fire truck with its sirens blaring approaches, we hear a higher-pitched tone than when it is receding.

---

[9] Max Karl Ernst Ludwig Planck (1858–1947), German Nobel-Prize-winning physicist whose discovery that energy exists in fundamental units called quanta marked the beginning of the development of quantum theory and the founding of modern physics. The Planck constant is equal to 6.6261×10^{-34} J s.

Figure 2.6 The displacement of a string away from its resting position: (a) as a function of time, (b) as a function of position.

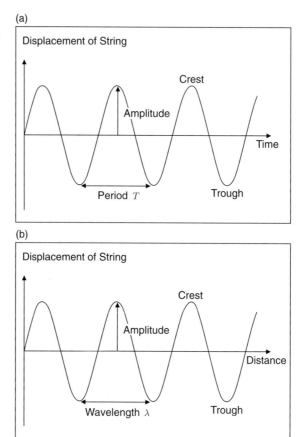

(a)

Displacement of String

Crest

Amplitude

Time

Period T

Trough

(b)

Displacement of String

Crest

Amplitude

Distance

Wavelength λ

Trough

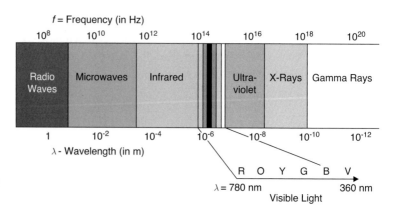

f = Frequency (in Hz)

| 10^8 | 10^{10} | 10^{12} | 10^{14} | 10^{16} | 10^{18} | 10^{20} |

| Radio Waves | Microwaves | Infrared | | Ultra-violet | X-Rays | Gamma Rays |

| 1 | 10^{-2} | 10^{-4} | 10^{-6} | 10^{-8} | 10^{-10} | 10^{-12} |

λ - Wavelength (in m)

R O Y G B V

λ = 780 nm   360 nm

Visible Light

Figure 2.7 The electromagnetic spectrum ordered left-to-right from low energy to high energy.

The effect only occurs for relative motion towards or away from the observer. The relationship between the component of velocity along a line connecting the emitter and the observer is

$$f' = f_0 \left(\frac{v \pm v_0}{v \pm v_s} \right),$$

where f' is the perceived frequency, f_0 is the actual frequency, v is the speed of the waves, v_s is the speed of the source (added to the wave speed if the source is moving away from the observer, subtracted if the source is moving toward the observer), and v_0 is the speed of the observer (added to the speed of the wave if moving toward the source, subtracted if moving away).

Photons and Energy

Light is one form of electromagnetic radiation. What is interesting about electromagnetic radiation is that it behaves as both a wave and a discrete particle. Light will act as a wave if we send it through a prism or diffraction grating and will separate out into a rainbow because of refraction. If we shine light at a metal, electrons can be emitted if the light's energy (or frequency) is high enough (the photoelectric effect). When the light energy is above a threshold level, the photoelectric effect demonstrates the particle nature of light (its frequency rather than its intensity determines whether electrons are emitted). Because EM radiation can act like a particle (i.e., it can be thought of as a discrete object as opposed to a continuous wave), it is given a special name – a photon. Therefore you can think of light as either a wave or a photon. How you think about it depends on the effect that you are observing. The bottom line, though, is that you should think of electromagnetic energy as waves or photons of energy. The amount of energy depends on the frequency of the radiation. This wave–particle duality of light caused lots of controversy early in the formation of modern physics and led to the development of quantum mechanics.

Blackbody Radiation

If you look closely at a candle flame, you will notice several colors in the vicinity of the wick (blue close to the wick and yellow and orange farther away). What we just learned about frequency (and hence color) tells us that the energy of the light we see must indicate different temperatures in various parts of the flame. All matter in the universe (including you) emits electromagnetic radiation. The amount and frequency of the radiation depends on the object's temperature. Your body temperature is normally near 98.6 °F (or 37 °C) and therefore you are emitting electromagnetic radiation mostly in the infrared (IR) portion of

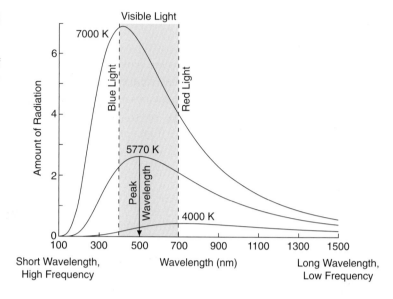

Figure 2.8 All objects emit EM radiation at a spectrum of wavelengths depending on their temperature. The peak of the blackbody curve moves to shorter wavelengths (higher energy) as the temperature increases. The Sun emits a spectrum corresponding to a photospheric temperature of about 5770 K.

the electromagnetic spectrum. Some animals such as nocturnal snakes have photoreceptors (a fancy name for eyes) sensitive to IR radiation. A candle flame burns hotter than our bodies' temperature, and therefore the radiation that it emits is higher energy (and frequency), so we see the light shining from a candle flame. This type of radiation is called blackbody radiation. Blackbody radiation has a property that makes it easy to identify – that is, the amount of radiation at any specific frequency depends only on the temperature. Figure 2.8 shows what is called a blackbody curve. The plot shows the amount of radiation emitted as a function of wavelength. Each curve represents an object at a certain temperature. Note that, as the temperature of an object increases, the amount of radiation emitted increases and the wavelength of the peak (or maximum) energy gets shorter. Since wavelength is inversely proportional to frequency, the peak frequency (and hence energy) emitted increases. A simple relationship called Wien's law[10] relates the peak wavelength to the temperature

$$\lambda_{\text{peak}} T = 2.898 \times 10^{-3} \text{m K}.$$

This law has enabled space scientists to work out the surface temperature of the Sun. We can measure the amount of electromagnetic radiation emitted from the Sun as a function of wavelength and then fit

[10] Wilhelm Carl Werner Otto Fritz Franz Wien (1864–1928), German Nobel-Prize-winning physicist whose work in thermodynamics and blackbody radiation led to the development of the relationship between wavelength and temperature for a blackbody radiator, now named Wien's law in his honor.

the data to a blackbody curve that has a nice analytical mathematical expression, and voila, we know its surface temperature. Much of astronomy and space science uses this relationship between the amount of radiation emitted as a function of frequency and temperature to understand the evolution and dynamics of stars and the temperature of planets. Note from Figure 2.8 that the Sun emits the most radiation in the visible part of the electromagnetic spectrum. Hence most animal eyes have evolved to be sensitive to visible light, while some nocturnal animals developed eyes sensitive to IR to see "cooler" objects at night.

2.8.3 Concept Mapping

A learning technique that is often discussed with active reading is called concept mapping. This learning technique involves creating a diagram that identifies, organizes, and connects concepts, processes, regions, and phenomenology together in a graphical representation. This helps you visualize the hierarchy of ideas and see connections between different concepts. The process of creating the diagram also helps identify the "big picture" and how subtopics are connected to the main topic. A concept map is often drawn with boxes for concepts and labeled arrows showing connections to other concepts. Creating concept maps helps with your own self-assessment of understanding, allows you to see growth in your understanding through the semesters, and is a powerful study tool. Figure 2.9 is an example of a concept map describing the "solar cycle." Usually the main concept is placed in the middle or at the top and different topics are connected hierarchically or radially from this main concept. The arrows or lines are often labeled with "active" verbs to help see the connections. Although you can use different shapes (boxes, circles) or lines (arrows or straight) to signify types of information or connections, the concept map should be simple to create and read. The concept map can be a study guide for future

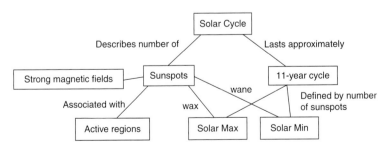

Figure 2.9 A concept map of the "solar cycle" showing how this tool can be used to organize new ideas, terms, and concepts together graphically. Note that the boxes are the "concepts," connected with descriptors to a definition or another topic, or may include a new concept with the definition.

reference, and different versions can be saved adding new information as introduced to explicitly connect the new information to your existing knowledge.

2.9 Problems

2.1 How many Earths could fit inside the Sun?

2.2 What has been the average length of time between solar maxima for the last five solar cycles (Figure 2.5)?

2.3 During fusion of hydrogen to helium, 4 billion kg of matter are converted to energy each second. What fraction of the Sun's total mass is lost each year to this process?

2.4 The Sun expels 1 billion kg of matter each second as the solar wind. What fraction of the Sun's total mass is lost each year to this process?

2.5 Order the following electromagnetic bands in terms of energy from lowest to highest: (X-ray, visible, gamma ray, radio wave, microwave, ultraviolet, infrared). Which band has the highest frequency? Which band has the longest wavelength? Which frequency of EM radiation corresponds to a wavelength of 1 m?

2.6 Using dimensional analysis (a technique of making sure both sides of an equation have the same physical dimensions), write a proportionality equation showing the relationship of wavelength to frequency (note that the product is equal to a constant – the velocity of the wave).

2.7 What is the thermal pressure of the gas at the surface of the Sun?

2.8 There is a direct relationship between an object's angular size and its distance from the observer. The Sun is 400 times the size of the Moon. How much farther away from Earth is the Sun than the Moon for them to have the same (0.5°) angular size as seen from Earth?

2.9 Using Wien's law, work out at what wavelength the Sun emits the most electromagnetic radiation. Is this in the visible part of the electromagnetic spectrum? If so, to what color does this correspond?

2.10 What is the Sun's total luminosity (power or energy per second)? What is the amount of power that intersects one square meter at the Earth? (This is called the solar constant.) At Mars? (1.5 AU)

2.11 What is the wavelength shift of a chromospheric spicule emitting in the $H\alpha$ that has a motion of 100 km s^{-1} towards the observer?

2.12 Create a concept map of sunspots connecting them to two other concepts discussed in this chapter.

2.13 Explain why sunspots have a leading and trailing magnetic polarity behavior and why they are opposite in the northern and southern solar hemispheres. Why does this polarity behavior flip every 11 years? Use Figure 2.4 in your answer.

2.14 If a sunspot has a temperature of 4000 K compared with the photosphere temperature of about 5800 K, what is the change in peak wavelength emitted between the two regions. What color does the sunspot's peak wavelength correspond to? Can you explain why sunspots appear black instead of closer to the color associated with their peak wavelength? (Hint: instead of color, think intensity. What would you see if you had a regular flashlight held up in front of a spotlight?)

Chapter 3
The Heliosphere

The heliosphere is defined as the region of interplanetary space where the solar wind is flowing supersonically.

(Dessler, 1967, p. 3: the first use of the term heliosphere in the scientific literature)

3.1 Key Concepts

- coronal mass ejection
- cosmic rays
- heliosphere
- interplanetary magnetic field
- plasma
- solar wind

3.2 Learning Objectives

After actively reading this chapter, readers will be able to:

- describe the regions, components, and events that exist in the heliosphere;
- distinguish between the direction of motion of the solar wind and the direction of the interplanetary magnetic field;
- assess the space weather impacts of coronal mass ejections and compare their impacts on the Earth's space environment with cosmic rays.

3.3 Introduction

Sunlight, which bathes Earth with heat and light, is only part of the energy that flows constantly from the Sun. Ionized gas (**plasma**) and magnetic field are continuously expelled as **solar wind**, as well. Solar wind was discovered in the 1950s when it was noticed that the plasma tail of a comet always points away from the Sun, even when the comet is moving back towards deep space. Figure 3.1 shows a schematic of a typical comet and its tail at different points in its orbit around the

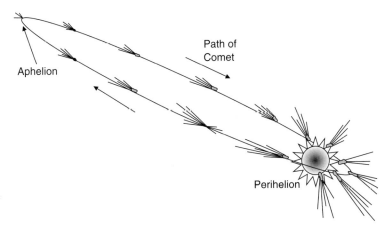

Figure 3.1 As a comet orbits the Sun, its tail always points away from the Sun whether it is moving towards or away from the inner solar system. Biermann used this observation to predict the existence of the solar wind, which continuously blows away from the Sun.

Sun. A comet tail is made up of cometary material that has been heated by sunlight as it gets close to the Sun and has escaped from the nucleus. The glowing cloud of neutral and ionized gas and dust around the nucleus is called the coma. The material in the coma is then "blown" back away from the Sun. Creation of a comet tail requires not only sunlight, but also the energy and momentum of a gas flowing supersonically away from the Sun, the solar wind. The solar wind expands out into interplanetary space and carves out a region surrounding the Sun called the **heliosphere**.

3.4 The Corona and the Solar Wind

During a total solar eclipse (when the Moon is lined up exactly between the Sun and Earth), light can be seen around the Sun. This light, referred to as the corona, is sunlight scattered by electrons in the Sun's outer atmosphere. The corona is not spherically symmetric or equally bright in all directions. Its spatial structure, with more emission (and hence plasma) near the equator than the poles, is due to the structure of the Sun's magnetic field. Plate 4 shows the corona during a total solar eclipse. The visible photosphere is a million times brighter than the corona, and therefore the corona can be seen only when photospheric light is blocked. Telescopes called coronagraphs, and those that can observe UV (see, e.g., Plate 3) and X-rays, now routinely observe the corona, even without a total solar eclipse.

For a particle to escape the Sun's gravitational pull in the lower atmosphere of the Sun, it must be moving faster than the Sun's escape velocity and be in a region tenuous enough that collisions with other particles are infrequent. The Sun's escape velocity is the speed at which a particle needs to be moving away from the Sun in order for it to never slow down, turn around, and fall back onto the Sun. The Sun's escape velocity is 618 km s^{-1} (the Earth's is 11.2 km s^{-1}). Coronal particles have a temperature of over 1 million kelvin, and feel pressure and electric forces giving velocities great enough to escape the Sun. These particles make up the solar wind, a plasma composed of mostly protons, helium nuclei, and electrons that moves supersonically away from the Sun and carries with it the Sun's magnetic field. Therefore the solar wind is a magnetized plasma.

3.5 Interplanetary Magnetic Field

The part of the Sun's magnetic field that is pulled out into the heliosphere by the solar wind is called the **interplanetary magnetic field** (IMF). Its characteristic spiral configuration when viewed from above or below the equatorial plane is due to the Sun's rotation. The magnetized solar wind expands radially (directly away from the Sun), pulling the solar magnetic field along with it. As the Sun rotates, the position or footpoint of where the solar wind stream leaves its surface moves counter-clockwise when viewed from above the Sun. This causes the magnetic field to start to spiral as it moves farther from the Sun with respect to the original footpoint position (see Figure 3.2). This is called an Archimedean spiral after the Greek scientist Archimedes,[1] who first described it mathematically. It is analogous to a stream of water being shot out of a rotating sprinkler head. Even though each individual drop or parcel of water moves radially away, since the sprinkler is rotating, the stream appears to curve or spiral out.

Because the Sun's rotation rate is essentially constant in time, the angle the spiral makes with respect to the Sun–Earth line (an imaginary line connecting the Sun and Earth) is due to the speed of the solar wind alone. Faster solar wind will create a smaller angle because it will move farther away from the Sun in a given time than a slower parcel. Because of the structure of the Sun's magnetic field (as revealed in the eclipse picture, Plate 4), the solar wind streams away from the Sun at different velocities. Solar wind flowing from closed-field regions near the Sun's equatorial plane is slower than that moving away from the Sun's open magnetic field regions. In the coronagraph photograph, regions of dipole-like field with a loop of plasma coming out of the limb are called closed-field regions. Near the poles, where a "ray" of plasma flows along a field line not looping back to the surface are called open-field

---

[1] Archimedes (c. 287–212 BC), Greek mathematician and scientist who is credited with discoveries in hydrostatics and mechanics. He is most famous for his Archimedes principle, which states that a body immersed in water will displace a volume of fluid that weighs as much as the body would weigh in air.

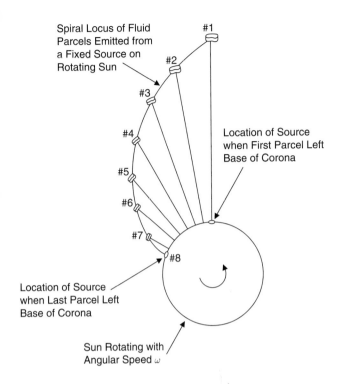

Spiral Locus of Fluid
Parcels Emitted from
a Fixed Source on
Rotating Sun

#1
#2
#3
#4
#5
#6
#7
#8

Location of Source
when First Parcel Left
Base of Corona

Location of Source
when Last Parcel Left
Base of Corona

Sun Rotating with
Angular Speed ω

Figure 3.2 Because of the Sun's rotation, its magnetic field is wrapped into an Archimedean spiral. The figure shows the location of eight parcels emitted at constant speed from a source fixed on the rotating Sun. (Source: Kivelson and Russell, 1995)

regions. In UV and X-ray photographs of the Sun (e.g., Plate 3), these open-field regions, referred to as coronal holes, appear dark. The location and existence of these two types of field regions change, which means that the speed of the solar wind observed at Earth changes constantly, but is most simply characterized by intervals of high-speed solar wind and slow-speed solar wind.

As a parcel of solar wind flows radially away from the Sun and hence the IMF is pulled farther away from the Sun, the angle of the Archimedean spiral (often called the Parker spiral when referring to the IMF, in honor of Eugene Parker, who first formulated the solar wind and IMF theory) becomes more and more azimuthal (the direction around the Sun). In other words, the IMF becomes less and less radial and more and more wrapped (perpendicular or orthogonal to the radial direction). So while the nominal Parker spiral angle at Earth is 45° (about half radial and half azimuthal), the IMF becomes near 90° (nearly azimuthal) as it moves out past Jupiter's orbit at 5 AU. As high- and slow-speed regions interact or run into and away from each other (called interaction regions), the solar wind and IMF compress or rarify, adding structure to the overall configuration.

The IMF has not only an Archimedean spiral pattern, but also structure in the north–south direction because the magnetic equator of

Figure 3.3 A perspective plot of the heliospheric current sheet showing the wavy nature of the interplanetary magnetic field due to the tilt of the Sun's magnetic axis with respect to its spin axis. (Source: Jokipii and Thomas, 1981)

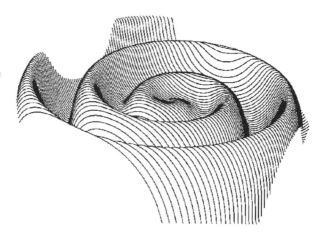

the Sun is not perfectly aligned with the spin axis of the Sun. Therefore, from the vantage point of Earth, we see solar wind coming from alternating sides of the Sun's magnetic equator as the Sun rotates. This gives rise to a flapping or wavy structure of the IMF. Figure 3.3 shows the structure of the IMF in three dimensions. The surface shown is the location of the plasma that came from the Sun's magnetic equator. This is called the heliospheric current sheet and divides IMF that is pointing away from the Sun from IMF pointing towards the Sun. There is also latitudinal (north–south) structure to the Parker spiral IMF, with the spiral becoming more wound (or azimuthal) with latitude due to the differential rotation of the Sun. With slower rotation near the poles, the footpoint on the surface of the Sun moves slower as the parcel of solar wind flows outward.

Over the 11-year solar cycle, the Sun's magnetic field structure changes as described in Section 2.7.3. The change from more ordered and dipolar during solar minimum to highly structured during solar maximum manifests itself in the structure of the IMF as well. During solar maximum, the IMF throughout the heliosphere is much more structured than during the more simple and ordered solar minimum configuration.

Table 3.1 gives some properties of the solar wind and the IMF.

3.6 Coronal Mass Ejections

The motion and structure of the Sun's magnetic field make the corona ever changing. Occasionally, reconfigurations of the solar magnetic field cause a large portion of the corona to blast away from the Sun

Table 3.1 *Average properties of the solar wind and IMF at 1 AU*

Number density	5 particles cm^{-3}
Temperature	1 000 000 K
Velocity	400 km s^{-1}
Composition	90% H, 8% He, and trace amounts of other heavy ions
IMF strength	10 nT

and out into the heliosphere. These **coronal mass ejections** (CMEs) were discovered with the first space-based coronagraph images in the 1980s. CMEs are large-scale magnetic structures that can contain over a trillion (10^{12}) kilograms of hot coronal material. A trillion kilograms is equivalent in mass to over a quarter of a million (250 000) aircraft carriers or a good-sized mountain. CMEs can move away faster than the background solar wind and have velocities of over 1000 km s^{-1} (several million miles per hour). Many of these fast CMEs set up shock waves in front of them as they stream away from the Sun. A shock wave is formed when the speed of an object exceeds the sound speed of the background material. Sound speed is dependent on a material's density and temperature. An example would be a bullet moving through air. The sound speed of air is approximately 300 m s^{-1}, so if the bullet moves faster than 300 m s^{-1}, a shock wave forms in front of it. With respect to a CME, the background solar wind is moving on average 400 km s^{-1}. The sound speed of the solar wind is approximately 40 km s^{-1}. Therefore, if a CME is launched with a speed greater than about 40 km s^{-1} faster than the background solar wind, a shock wave will form. One important aspect of a shock wave in interplanetary space is that it is a very good particle accelerator. Therefore fast CMEs drive a shock that can generate large numbers of energetic particles as it plows through the slower solar wind. These solar energetic particles can reach Earth and cause damage to satellites.

CMEs often have distinctive, loop-like magnetic field structures called flux ropes (see Plate 2, an image from the NASA SOHO satellite). When referring specifically to a CME, they are called magnetic clouds. When observed in interplanetary space, CMEs have distinct plasma and magnetic field properties that are used to identify them. They often take over a day to pass by Earth, implying a length scale of a quarter of an astronomical unit (0.25 AU). CMEs often have a shock and high-density "plug" of plasma in front due to slower solar wind plasma being "swept up" like snow by a snowplow. Fast CMEs are the leading cause of major geomagnetic storms and are therefore one of the most important solar phenomena that influence space weather.

3.7 The Outer Heliosphere

The outer heliosphere is defined as the region well beyond the orbit of Pluto where the solar wind interacts directly with interstellar space. Pluto is on average 40 AU away from the Sun. The heliopause – the boundary between the heliosphere and the interstellar medium (ISM) – is about 120 AU away. In 2003 Voyager I passed the termination shock (formed between the supersonic flow of the solar wind and the interstellar medium) when it was 90 AU from the Sun, and in 2012 it passed the heliopause when it was about 122 AU from the Sun. Voyager 2 passed the heliopause in December 2018 when it was about 120 AU from the Sun. The ISM is a mixture of electrically neutral atoms and magnetized plasma. Since the Sun orbits the center of our galaxy at approximately 220 km s^{-1}, there is a relative velocity difference between the heliosphere and the surrounding ISM. This gives rise to a teardrop-shaped heliosphere with a termination shock inside the heliopause (see Plate 5).

The heliopause proper separates the Sun's magnetized plasma from the ISM. This is analogous to the magnetopause that separates Earth's magnetic field from the magnetized plasma of the shocked solar wind. In the direction of motion of the Sun, another bow wave or shock is formed between the relatively moving heliosphere and the ISM. The heliosphere is estimated to be moving roughly perpendicular to the line between the Sun and the center of the Milky Way galaxy – hence we are approximately orbiting the galaxy in a circular path. At 220 km s^{-1} it takes approximately 250 million (250 000 000) years to orbit the galaxy once. In its 4.5 billion-year history, Earth has orbited the galaxy 18 times. So Earth is about 18 "galactic" years old. Beyond this bow wave or shock is unperturbed ISM that extends out to the next stellar atmospheres. The closest star, Proxima Centauri in the Alpha Centauri star system, is about 4 light-years (LY) away (a light-year is the distance light travels in one year, about 10 trillion km or 60 000 AU). Of course, there is space weather around that star as well. In 2016 an exoplanet was discovered orbiting Proxima Centauri in the so-called habitable zone – the region not too close and not too far from the star for liquid water to potentially exist on the surface. However, the solar wind pressure is expected to be thousands of times stronger than at Earth, so it is not known if conditions on the surface would be hospitable for life.

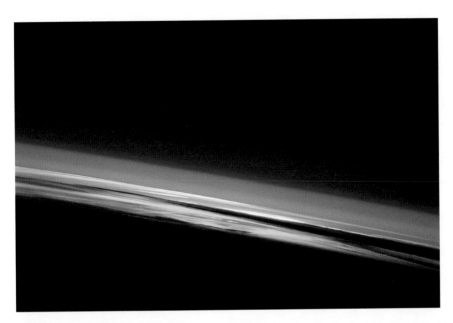

Plate 1. The limb of Earth taken from the International Space Station. Notice the sharp edge to the blue of the atmosphere against the black of space. (Credit: NASA)

Plate 2. A large coronal mass ejection (CME) ejects a cloud of particles into space on December 2, 2003. In this composite an EIT 304 image of the Sun from about the same time has been appropriately scaled and superimposed on a LASCO C2 image where a red occulting disk can be seen extending around the Sun. This LASCO coronagraph instrument allows details in the corona to be observed. (Courtesy of SOHO/LASCO consortium. SOHO is a project of international cooperation between ESA and NASA)

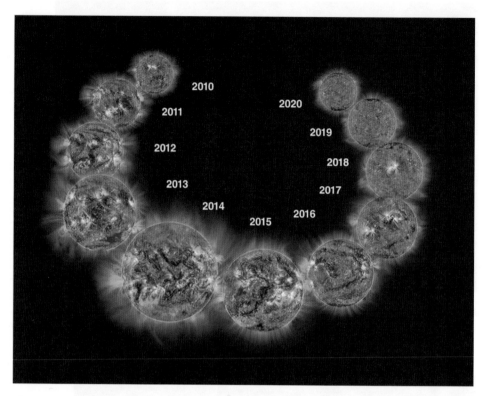

Plate 3. Evolution of the Sun in extreme ultraviolet light from 2010 through 2020, as seen from the telescope aboard Europe's PROBA2 spacecraft. (Credit: Dan Seaton/European Space Agency and collage by NOAA/JPL-Caltech)

Plate 4. During a total solar eclipse, the Moon passes directly in front of the Sun, completely blocking out the photosphere. The chromosphere and corona then become visible for the few minutes the eclipse lasts. This photograph was taken during the total solar eclipse on August 21, 2017. (Credit: NASA/Gopalswamy)

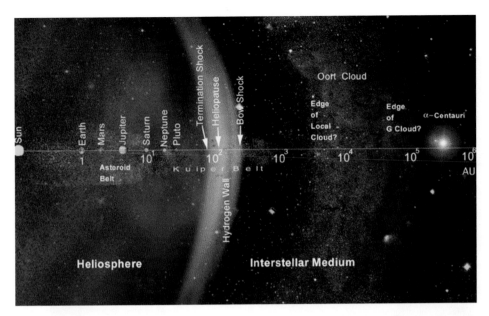

Plate 5. The heliosphere and local interstellar medium out to our nearest stellar neighbor – Alpha Centauri. Note that the scale is logarithmic in astronomical units (AU) with each tick mark 10 times farther from the Sun than the previous one. (Credit: NASA Interstellar Probe Science and Technology Definition Panel, 1999)

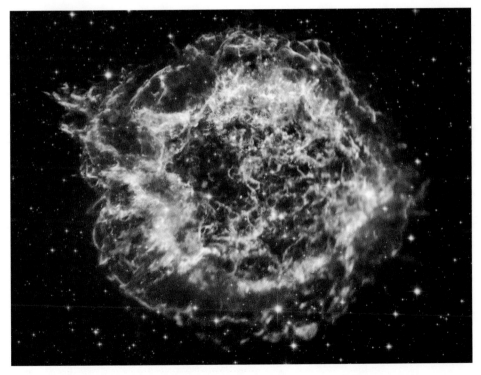

Plate 6. The supernova remnant Cassiopeia A, a composite image of visible light from Hubble with X-ray observations by Chandra. The different colors represent different energies of the light, which tend to come from different elements such as oxygen, calcium, and iron. (Credit: X-ray: NASA/CXC/SAO; optical: NASA/STScI)

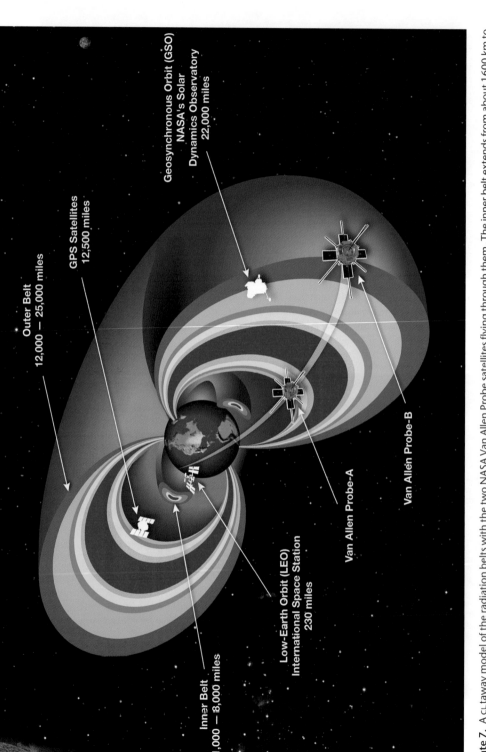

Outer Belt
12,000 – 25,000 miles

GPS Satellites
12,500 miles

Geosynchrronous Orbit (GSO)
NASA's Solar
Dynamics Observatory
22,000 miles

Van Allen Probe-A

Van Allén Probe-B

Low-Earth Orbit (LEO)
International Space Station
230 miles

Inner Belt
1,000 – 8,000 miles

Plate 7. A cutaway model of the radiation belts with the two NASA Van Allen Probe satellites flying through them. The inner belt extends from about 1600 km to 12 800 km (1000–8000 miles) above Earth's equator. The outer belt extends from about 19 200 km to 40 000 km (12 000–25 000 miles). This graphic also shows other satellites near the region of trapped radiation. (Credit: NASA)

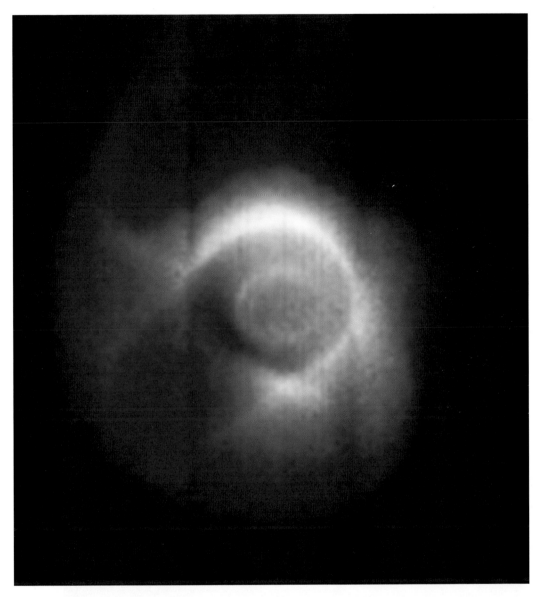

Plate 8. The plasmasphere as observed in extreme ultraviolet light from the IMAGE spacecraft. The blue is a false color image of the high-density plasmasphere that extends several Earth radii from the surface of Earth. The sharp density boundary called the plasmapause is clearly visible in this image. The view is from over the north polar cap, with the Sun to the upper right corner. The auroral oval is visible at the center of the image. (Credit: IMAGE Mission, courtesy of Bill Sandel)

Plate 9. The northern lights, as seen in the sky over Alaska on the night of February 16, 2017, from the Poker Flat Research Range north of Fairbanks. (Credit: NASA/Terry Zaperach)

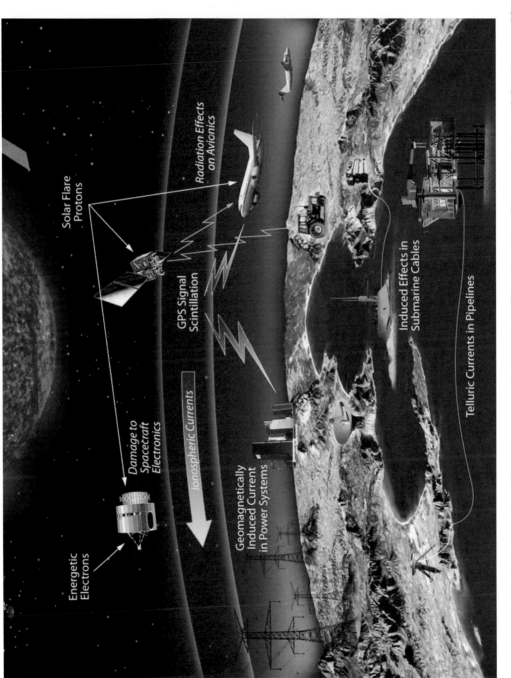

Plate 10. Different technological systems affected by space weather include satellites, astronauts, radio communication, and electric power grids. (Credit: NASA)

Plate 11. A closeup picture of one of the transformers melted by the electrical overload brought about by a geomagnetic storm. (Credit: CT Gaunt/NCU. The source of this material is the COMET® website at http://meted.ucar.edu/ of the University Corporation for Atmospheric Research (UCAR), sponsored in part through cooperative agreement(s) with the National Oceanic and Atmospheric Administration (NOAA), US Department of Commerce (DOC). © 1997–2021 University Corporation for Atmospheric Research. All Rights Reserved)

Plate 12. Meteor crater in northern Arizona formed when an asteroid 50 meters in diameter hit about 25 000 years ago. (Credit: image taken June 7, 2007, courtesy of NASA)

3.8 Cosmic Rays

Earth is constantly bombarded from every direction by highly ionized atoms and other subatomic particles known as **cosmic rays**. "Ray" is a misnomer, since cosmic rays consist of energetic particles. Cosmic rays are further divided into two components: particles that originate outside our heliosphere (called galactic cosmic rays) and those that originate from our Sun (called solar energetic particles). Cosmic rays travel at nearly the speed of light, and most are nuclei of atoms. The composition of cosmic rays spans the periodic table, from the lightest particles, such as hydrogen and helium, to the heaviest, such as iron. Cosmic rays also include electrons, positrons (the mirror particle of an electron — essentially a positively charged electron), and other sub-atomic particles. The energy of cosmic rays is usually measured in MeV (mega-electronvolts or a million eV) or GeV (giga-electron-volts or a billion eV). An electronvolt is an energy unit equivalent to the energy gained by an electron accelerated through a one-volt poten-tial electric field. One eV is equal to 1.6×10^{-19} joules (J). Typical energies of galactic cosmic rays are between 100 MeV and 10 GeV. To put this in perspective, if the galactic cosmic ray is a proton, it has a velocity between 43% and 99.6% the speed of light to have this much energy. The highest-energy cosmic rays ever measured have kinetic energies equivalent to that of a baseball thrown by a professional pitcher – all contained in a single atomic nucleus.

The current theory is that most galactic cosmic rays originate in supernovas (stellar explosions). It is estimated that one supernova happens every 50 years in a galaxy like the Milky Way. One type of supernova is the death-throes of a massive star. After the star has consumed all of its fuel by thermonuclear fusion, its outer layers collapse and cause a huge explosion that expels stellar material into space and causes shock waves to form. The explosion and shock waves then produce particles of very high energy. The shock waves continue to propagate away from the progenitor star (the star that became a supernova), continuously accelerating particles for many years after the explosion. Plate 6 shows a picture of a supernova remnant Cassiopeia A. The object is about 11 000 LY away from us and it is estimated that the first light of the supernova reached Earth about 350 years ago, though no one at the time noticed or recorded the event. As the shock wave continues to expand, particles can be continuously accelerated giving rise to a bath of cosmic rays throughout the galaxy.

Since cosmic rays are charged particles, their motion is deflected by galactic magnetic fields as they propagate through interstellar space. Therefore it is not possible to identify directly their source since their path is a "random walk" from their source to us (i.e., since the galactic magnetic field is highly structured, cosmic rays are essentially scattered in every direction as they propagate through space). The Sun's magnetic field (the IMF) and Earth's magnetic field also influence the motion of cosmic rays through our solar system and to Earth's surface. Because the IMF's structure changes from ordered to highly structured from solar minimum to solar maximum, the flux of galactic cosmic rays waxes and wanes from solar minimum to solar maximum. In other words, the flux of galactic cosmic rays peaks during solar minimum and is least at solar maximum. This is out of phase with solar energetic particle fluxes that peak during solar maximum.

When high-energy cosmic rays hit Earth's atmosphere, they collide with atmospheric particles and cause showers of secondary particles to hit the surface. Each collision takes energy from the original cosmic ray, creates new particles, and energizes the atmospheric gas particles that are hit. These in turn can hit other particles, energizing them, which can then hit new particles, and so on. Depending on the energy of the incoming cosmic ray, large fluxes of secondary particles can reach Earth. A by-product of these collisions is the creation of pions, unusual subatomic particles that usually decay quickly to produce muons, neutrinos, and gamma rays. Muons also subsequently decay into electrons and positrons. The flux of these particles is equivalent to about 1000 particles per minute passing through our bodies. However, the effect is only a small part of the natural background radiation. In space the flux can be considerably higher and cause damage (or death) to astronauts and satellites.

3.9 Supplements

> Can no one laugh? Will no one drink? I'll teach you physics in a wink.
>
> (Gamow, 1972, p. 190)

3.9.1 Additional Learning Objectives

After actively reading these supplements, readers will be able to:

- calculate the velocity, acceleration, and force of a moving object given its position as a function of time and use this information to predict the future location of a moving object;
- classify types of knowledge using Bloom's Taxonomy, use that information to build conceptual understanding and frameworks of new knowledge, and use the new knowledge to solve problems and evaluate models.

3.9.2 How Do We Describe Motion?

Scientists and engineers study the motion of everything from golf balls, satellites, blood cells through the body, chemicals through the atmosphere, to radiation in space, cars involved in a collision, and Earth around the Sun, just to name a few diverse examples. In order to understand motion of an object, several things need to be defined, such as what does it mean to move?

Mechanics is the study of moving bodies or objects. Motion is defined as the change in position of a body with respect to another body or to some reference frame (such as a room or Earth). It is often convenient to define a fixed reference frame (called an inertial reference frame) so that observers can describe to each other the motion of any object in a way that makes sense to both. The development of the concept of a reference frame is one of the great achievements of math and science. For example, you are probably reading this chapter sitting down. You would describe your motion as zero, or in other words, you are not moving with respect to your surroundings. This is true if you define your surroundings as the room. However, what if you are sitting in the back seat of a car driving down a road? You are not moving relative to the driver, other passengers, or the car's interior, but to an observer on the side of the road, you have motion relative to the road and its surroundings. Therefore, when you describe motion (or lack of motion), you must also specify the reference frame you are using. A fundamental principle of physics is that the laws of physics are valid and identical in all reference frames. For space physics, the frame of reference is often the fixed Earth or the Sun. For example, we can say that we are sitting still with respect to the surface of Earth, but an observer fixed in space directly overhead will see us move towards the east as Earth rotates. An observer fixed with respect to the Sun would see not only us rotating with Earth, but also Earth receding as it orbits the Sun. An observer fixed with respect to the center of the Milky Way galaxy would see us rotating around the Earth's axis, orbiting the Sun, and the Sun in turn moving around the center of the galaxy. Of course we can move farther out and use a system fixed not to our galaxy, but to the relative positions of the different galaxies in our local neighborhood (though, to make things more complicated, those galaxies are also moving, so sometimes it is difficult to clearly define where you are or even if you are moving). Any observer can easily transform their calculations from one reference frame to another, though in each succeeding step we need more information about the relative motion and position of each reference frame to the succeeding reference frame.

In order to describe the position of an object in a reference frame, we must define a coordinate system. A coordinate system is a set of

Figure 3.4 A two-dimensional Cartesian coordinate system.

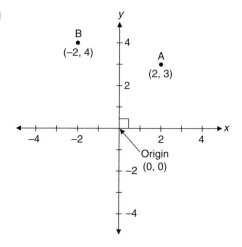

rules that describes quantitatively where an object is located from a specific point, called the "origin." Figure 3.4 shows a two-dimensional coordinate system you are probably familiar with. The x- and y-axes form a right angle (90°). This system, first developed by French mathematician René Descartes, is called a Cartesian coordinate system in his honor. This coordinate system can be used to describe the position of an object in a plane. A plane is a two-dimensional region (like the surface of a piece of paper) that can be described by two coordinates (in the example, x and y). A three-dimensional coordinate system would need a third axis (usually denoted z) to specify the position above or below the x–y plane. The tick marks on the axis can be used to describe a point anywhere on the coordinate system. The origin is defined as being at position $(0, 0)$ and by convention we define distances to the right and up as increasing positive numbers and those to the left and down as increasing negative numbers. Point A in Figure 3.4 is therefore at position $(2, 3)$; in other words, 2 tick marks over to the right of the origin on the x-axis and 3 tick marks up the y-axis. Point B in Figure 3.4 is at position $(-2, 4)$ or 2 tick marks to the left of the origin and 4 tick marks up the y-axis. Street maps are usually drawn on a similar coordinate system, with the axes labeled North, South, East, and West.

If we want to understand the motion of a parcel of the solar wind on its trajectory to Earth, we often use a coordinate system centered on the Sun. One such coordinate system is the solar ecliptic frame of reference. The center of the Sun is at the center of this system (or at the origin), and a line connecting the center of the Sun and the center of Earth is defined as the x-axis. The z-axis is the line perpendicular (or at a right angle) to this line and to the plane that contains Earth in its yearly orbit around the Sun. You can visualize this system as a vinyl record, CD, DVD, or Blu-ray disk. The Sun is at the center of the disk and Earth travels about the Sun on the surface of the disk. The two-dimensional plane that contains the center of the Sun and Earth as it sweeps out its motion around the Sun is called the ecliptic plane. This plane also contains all the planets in the solar system. We can now say this without the Pluto caveat. Pluto's orbit is inclined to the ecliptic plane by 17°. This large inclination, which is different from those of the other eight planets, contributed to Pluto's demotion to dwarf planet status in 2006.[2]

If we are interested in locating a place on the surface of Earth, we usually use a geographic coordinate system with degrees instead of distance units because Earth can be represented as a sphere or globe. In this system, we define the origin as the point at the equator where the meridian (the north–south line that crosses the equator at right angles) goes through London. This has been defined as the 0 (or prime) meridian (the British Navy ruled the seas in the seventeenth and eighteenth centuries and hence were able to define the meridian that contained their main observatory at Greenwich as the reference line, which is still in use today). Since there are 360° around a circle, the globe is marked off in degrees starting at the 0 meridian and going eastward all the way around Earth. This coordinate is called longitude. The position above or below the equator is also given in degrees. By convention we define the North Pole to be +90° and the South Pole to be –90°. Do you know your present latitude and longitude?

Velocity

Once we have defined a reference frame and a coordinate system, we can measure an object's change in position as a function of time. We can note the position of a car driving down a highway (designated as x_1) at some time (designated t_1) and then measure its position (x_2) at some later time (t_2). We can then say that the car moved distance ($x_2 - x_1$) in ($t_2 - t_1$) amount of time. The ratio of these differences is called a speed:

$$\text{Speed} = \frac{(x_2 - x_1)}{(t_2 - t_1)} = \frac{\Delta x}{\Delta t}.$$

---

[2] In 2006, Pluto was reclassified as a dwarf planet. This change was spurred by the discovery of another object larger than Pluto called Eris amongst thousands of other bodies that share orbital characteristics with Pluto. These bodies are called Kuiper Belt objects. In 2015, New Horizons flew by Pluto taking the first high-resolution images of its surface.

Speed has the dimensions of distance divided by time, or in standard metric units (Système International, SI) meters per second. In our cars we usually denote speed in miles per hour or kilometers per hour. Speed is an important parameter for understanding the dynamics of an object, but one additional parameter, the direction of motion, is needed for us to predict the future position of an object. We define an object's speed and direction of motion as its velocity. Velocity is a vector – a mathematical quantity that has both a magnitude (speed) and direction. Speed is a scalar (a parameter with a magnitude, but no direction). Vector mathematics are the rules that are followed when describing vector quantities. The simplest vector mathematics that we can do is adding two vectors. As an example, let us determine the velocity of a person walking down the aisle of a moving train. If the train is moving at 10 km per hour due north and the person is walking north towards the front of the train at 2 km per hour, then the total velocity of the person relative to the stationary tracks is ($\vec{v}_{train} + \vec{v}_{person}$) or 12 km per hour heading north. Note the arrows above the velocity symbol. These arrows signify that the quantity is a vector rather than a scalar quantity. Now consider a person walking south to the back of the train (the velocity of the train and the velocity of the person on the train are in opposite directions). If we define moving north as positive, then moving south would be described as having a negative velocity. Therefore the resultant sum of the velocity vectors of the train moving north at 10 km per hour and the passenger walking toward the back of the train (south) at 2 km per hour would be $(10 - 2)$ or 8 km per hour northward. Note that the person is still moving north relative to the tracks even though he is moving south inside the train because the train is moving faster than the passenger is walking. This is similar to walking up an escalator. Your resulting speed is the sum of the escalator's speed and your walking speed. Now if you attempt to walk up a "downgoing" escalator you can still go up as long as you walk faster than the escalator is traveling down. However, your speed relative to the building will be less than your speed relative to the escalator steps.

In space physics we are often interested in the velocity of a parcel of plasma from the Sun on its way to Earth. In the solar ecliptic reference frame, the motion of the solar wind is radially outward from the Sun. Recall that in 1859 Richard Carrington observed a geomagnetic effect that he suggested might be associated with a solar flare observed 17 hours earlier. The average speed of the plasma between the Sun and Earth can be calculated by looking at the ratio of the distance traveled to the time it takes to reach Earth ($v = d/t$). The Sun is 150 million km away from Earth, so the average speed the parcel of solar wind must have had if the solar flare and geomagnetic storm were related is: 150 000 000 km/17 hours = 2450 km s^{-1}. The 1859 storm has been re-examined and found to be the strongest geomagnetic storm on record. This velocity is one of the highest ever estimated for a parcel of solar wind, which has an average velocity of about 400 km s^{-1}.

Acceleration

To predict where an object will be in the future, we need to know not only its position at different times, but also its velocity at those times. By observing whether the velocity changes at each time step, we can learn if the object is slowing down, speeding up, or changing direction. If the object is doing any of these three things, we say that it is accelerating (or changing its velocity with time). Note that an object's speed can be constant, but if it is changing its direction then the velocity vector is changing. Therefore acceleration can be defined by comparing the difference in velocities at two different times. In a method similar to that used to calculate velocity, we can write this difference as

$$\vec{a} = \frac{(\vec{v}_2 - \vec{v}_1)}{(t_2 - t_1)} = \frac{\Delta \vec{v}}{\Delta t},$$

where the symbol "a" denotes acceleration. Note also that acceleration is a vector; it has a magnitude and direction. Acceleration tells us how much an object's velocity is changing as a function of time. If we observe an object's velocity increasing with time, we say it is accelerating. If it is decreasing, we often say it is decelerating.

Force

What makes an object speed up, slow down, or change direction? For objects we typically use – chairs, books, coffee mugs – we provide a push or pull that moves these objects. The formal name for an action that causes an object to change its velocity (or acceleration) is force. Sir Isaac Newton, one of the greatest mathematicians and scientists of all time, developed the laws of motion that accurately describe the motion of almost everything – from falling apples, to planets, to stars.

The "almost" is used because Einstein later found that if things move really fast – near the speed of light – then Newtonian mechanics or Newton's laws of motion need to be modified. This modification is called special relativity. Einstein also found that when describing motion near a massive object such as a star, Newtonian mechanics needs to be modified by general relativity. Finally, modern physics discovered that if you are describing the motion of subatomic objects like electrons, you need to use quantum mechanics instead of Newtonian mechanics.

For everything not moving near the speed of light, near a massive object or the subatomic – the use of Newtonian mechanics is appropriate. One of Newton's laws defines force. Force is equal to the mass of an object times its acceleration ($\vec{F} = m\vec{a}$). (Note the arrows above the force and acceleration variables. Force is a vector and therefore has magnitude and direction.) A force will accelerate a mass. One force we deal with every day of our lives is the gravitational force between the Earth and us. The mass of Earth "pulls" us down to the surface. Therefore, if I drop a pencil from my desk, it will fall straight down to the floor (the gravitational force vector points to the center of Earth). As the pencil falls, it accelerates continuously (going faster and faster) until it hits the ground. Near the surface of Earth, the magnitude of the gravitational acceleration is 9.8 m s^{-2}. This means that an object dropped from some height will accelerate (or speed up) by 9.8 m s^{-1} every second (neglecting air resistance). So after two seconds the object is moving 19.6 m s^{-1} (~70 km h^{-1}).

3.9.3 Bloom's Taxonomy

How do we learn? What do we do when we encounter new knowledge and how do we use that new knowledge with our previous understanding to make decisions, solve problems, explain our observations, and create new ideas?

Educators and cognitive scientists have developed a classification of learning goals or cognitive processes to help educators and learners to understand how we learn and use new knowledge. Originally developed by Benjamin Bloom and colleagues in the 1950s (hence Bloom's Taxonomy), it has been subsequently revised to help us see the different cognitive processes that we need to develop to move from novice to expert.

Bloom's Taxonomy is often represented as a pyramid (Figure 3.5) with "remembering or recalling" knowledge (facts, concepts, and processes) at the base to signify that the skills or processes higher in the pyramid all rely on our ability to read, understand, and recall knowledge for us to be effective thinkers. The definition of the different levels and sublevels that give examples of the type of processes within each level of the taxonomy are: remember (recognize, recall), understand (interpret, exemplify, classify, summarize, infer, compare, explain), apply (execute, implement), analyze (differentiate, organize, attribute), evaluate (check, critique), and create (generate, plan, produce).

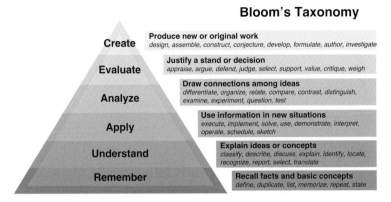

Bloom's Taxonomy

Create — Produce new or original work
design, assemble, construct, conjecture, develop, formulate, author, investigate

Evaluate — Justify a stand or decision
appraise, argue, defend, judge, select, support, value, critique, weigh

Analyze — Draw connections among ideas
differentiate, organize, relate, compare, contrast, distinguish, examine, experiment, question, test

Apply — Use information in new situations
execute, implement, solve, use, demonstrate, interpret, operate, schedule, sketch

Understand — Explain ideas or concepts
classify, describe, discuss, explain, identify, locate, recognize, report, select, translate

Remember — Recall facts and basic concepts
define, duplicate, list, memorize, repeat, state

Figure 3.5 Bloom's Taxonomy outlines the hierarchy of knowledge, showing how different types of knowledge use different processes to understand, solve problems, and create new knowledge. (Source: Vanderbilt University)

As you learn about space weather (or any new topic), understanding the types of knowledge and the different skills, abilities, and cognitive processes that help you organize, understand, and use knowledge enables more effective and efficient learning and improves your cognitive abilities, enabling you to move from novice to mastery.

Think back to the types of exams that you took in high school. Many of the problems were "remember or recall" type questions, and studying involved memorizing vocabulary lists, techniques (like how to diagram a sentence or how to determine the area of a triangle), and names and dates. In college and in the workforce, the expectation is that you develop higher-order thinking skills (those higher up the Bloom's Taxonomy pyramid) and can understand knowledge, apply it to new problems, analyze and synthesize the knowledge with your existing understanding to evaluate competing ideas and create new ideas. Note that the next level above recall and remember is "understand." How can you test yourself to see if you understand new knowledge (which is different than just remembering the definition of a new vocabulary term)? One way is to see if you can compare the new knowledge with existing knowledge or with other new concepts. These "compare and contrast" questions enable you to see similarities and differences that will then allow you to apply this knowledge to solve problems. For example, can you compare and contrast solar flares with coronal mass ejections? What are the similarities and differences? Once you understand this, can you apply this new knowledge to understand why the impacts of solar flares on Earth are essentially immediate and impulsive (short duration), while even though coronal mass ejections usually take a day or two to reach Earth, their impact can be felt prior to their arrival (due to the shock wave accelerating particles that can travel much faster than the CME itself), and once CMEs hit Earth their impact can be much more catastrophic in terms of space weather storms than flares?

3.10 Problems

3.1 Estimate the time required for a parcel of solar wind with a speed of 800 km s^{-1} to travel from the surface of the Sun to Earth. Is Carrington's observation of possible cause and effect of a solar flare and geomagnetic activity consistent with this time?

3.2 How long does it take electromagnetic radiation moving at the speed of light to reach Earth from the Sun?

3.3 The Parker spiral of the IMF depends on the speed of the solar wind.

(a) For a 400 km s^{-1} solar wind, what is the angle of the IMF with respect to the Sun–Earth line?

(b) In which direction does the solar wind move with respect to the Sun?

(c) In which direction does the solar wind move with respect to the Sun–Earth line? (Note: the Earth has an orbital velocity around the Sun of 30 km s^{-1}.)

3.4 Which parameters determine the shape and size of the heliosphere?

3.5 Compare and contrast CME with cosmic rays.

3.6 If a parcel of solar wind were radially accelerated at 10 km s^{-1} away from rest, how long would it take to be going faster than the escape velocity of the Sun?

3.7 If the ISM properties (plasma and magnetic field pressure) were relatively constant, describe how the size of the heliosphere would change over the 11-year solar cycle.

3.8 The New Horizons spacecraft made a flyby of Pluto in 2015 and is one of the fastest spaceships to ever explore the solar system. At 80 000 km h^{-1}, the approximate current speed of the New Horizons spacecraft, how many AU per year is it traveling? If it was traveling in the direction of Proxima Centauri, how many years would it take?

3.9 Using Bloom's Taxonomy, evaluate the level of questions 3.4 and 3.6 and describe their differences.

3.10 Write a series of questions about a topic discussed in Chapter 3 and answer them corresponding to the first three levels of Bloom's Taxonomy. Can you explain the difference in the way knowledge is used to answer the different questions?

Chapter 4
Earth's Space Environment

It has now become possible to investigate the region above the ionosphere in which the magnetic field of the earth has a dominant control over the motions of gas and fast charged particles ... it may appropriately be called the magnetosphere.

(Gold, 1959: from the paper that coined the term magnetosphere)

4.1 Key Concepts

- geomagnetic storm
- magnetic field
- magnetic reconnection
- magnetosphere
- Van Allen radiation belt

4.2 Learning Objectives

After actively reading this chapter, readers will be able to:

- name the regions of the Earth's magnetosphere and describe how the Earth's magnetic field defines its structure;
- explain how magnetic reconnection gives rise to geomagnetic storms and substorms that drive magnetospheric dynamics;
- predict what solar wind conditions will give rise to geomagnetic storms and changes in the Earth's ring current and radiation belts.

4.3 Introduction

At approximately 100 km (or about 60 miles) above Earth's surface, the amount of ionized gas becomes appreciable. Because ionized gas is electrically charged, it feels the effect of Earth's **magnetic field**, which plays an important role in guiding the motion of charged particles in near-Earth space. Through its interaction with the magnetized solar wind, Earth's magnetic field is intimately involved in coupling or transferring energy and momentum from the Sun into our space environment. This chapter describes the magnetic field region surrounding

Earth called the **magnetosphere**. The connection of the magnetosphere with the Sun is at the heart of space weather.

4.4 Dipole Magnetic Field

Magnetic fields are force fields around magnets, electric currents, or moving charged particles that exert a force on other magnets, electric currents, or moving charged particles. Due to the motion of molten iron inside Earth, a relatively strong magnetic field surrounds it.

Like the magnetic field in sunspot pairs or a refrigerator magnet, Earth's magnetic field emerges from one hemisphere with a certain direction and points towards the opposite hemisphere. In general, for magnetic fields emanating from things like refrigerator magnets, the north pole is defined as the pole where the magnetic field points outward and the south pole where it points inward. Imagine a magnet inside Earth with the north magnetic pole pointing south and the south magnetic pole pointing north (see Figure 4.1). This is the configuration of Earth's magnetic field. To avoid confusion between Earth's north geographic pole and its magnetic north pole, the magnetic north pole is defined to be in the northern hemisphere. This potentially confusing difference between Earth and a refrigerator magnet is done so that Earth's north magnetic pole is in the same hemisphere as its north geographic pole. We call the point where the magnetic field emerges straight out of Earth the south magnetic pole and the point where it goes directly into Earth the north magnetic pole. Such a magnetic field is referred to as a dipole ("di" being Greek for "two"). Earth's magnetic poles are not located in the same place as the geographic North and South Poles, which are defined by Earth's spin axis. The magnetic dipole axis is tilted by about 11° with respect to the spin axis. In Figure 4.1, the direction of each

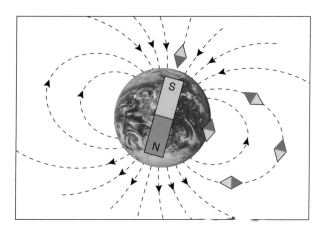

Figure 4.1 Earth has a dipole magnetic field with the same shape as that of a regular magnet. (Source: NASA StarGazers)

field line is indicated; the spacing of the lines represents the strength of the field. Note that the field is stronger at the poles than at the equator. The strength of a dipole magnetic field is two times greater at the magnetic pole than at the equator and falls off very quickly with distance; the strength of the field at the equator decreases as the cube of the distance:

$$\left(|B| \propto \left(\frac{1}{r^3} \right) \right).$$

Satellites with instruments that measure the strength and direction of magnetic fields have explored much of the space around Earth and all of the planets. These missions have verified the dipole nature of Earth and the other planets with magnetic fields (Mercury, Earth, Jupiter, Saturn, Uranus, Neptune).

4.5 Structure of the Inner Magnetosphere

Figure 4.2 shows a cross-section of the magnetosphere in the noon–midnight meridian, with north at the top and the Sun on the left. The regions of the magnetosphere are labeled. Note that the magnetic field resembles a dipole close to Earth. The dipole region of Earth's magnetosphere is called the inner magnetosphere. On the nightside at about geosynchronous orbit (6.6 Earth radii [r_E] from the center of Earth), the magnetic field lines become stretched into a long, tail-like configuration. The interaction of Earth's magnetic field with the solar wind is

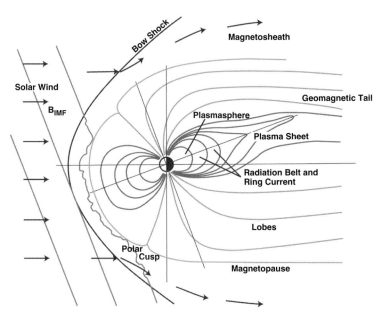

Figure 4.2 A noon–midnight cross-section of Earth's magnetosphere. Note the dipole shape of the inner magnetosphere. The Sun (hence noon) is to the left and north is up.

responsible for the distortion of its dipole field. The non-dipolar regions are called the outer magnetosphere.

Immediately surrounding the Earth is a region of cold (about 1 electronvolt, eV), dense (tens to thousands of particles per cm^3) plasma that essentially co-rotates with Earth. This region is called the plasmasphere. An electronvolt is a measure of kinetic energy. For a proton, 1 eV corresponds to a velocity of about 14 km s^{-1}. Densities in space are much lower than at Earth's surface. The density of air at sea level is approximately Avogadro's number.[1] In the densest region of the magnetosphere – the plasmasphere – densities are billions and billions of times lower.

The plasmasphere consists mostly of hydrogen and helium, but also an appreciable amount of oxygen that has just enough energy to escape from Earth's ionosphere. The ionosphere, created by solar ultraviolet and X-ray radiation, will be discussed in much more detail in Chapter 5. As plasma drifts up the magnetic field line from below, it becomes trapped and co-rotates with Earth. There is often a very sharp boundary to the dense plasmasphere called the plasmapause. Plasma density frequently drops an order of magnitude within a very short radial distance (less than 0.5 r_E).

Often overlapping with the plasmasphere are the **Van Allen[2] radiation belts** and the ring current. These two regions are characterized by high-energy particles that are magnetically trapped in Earth's magnetosphere. The ring current is made up of particles with a peak energy of about 200 keV, while the radiation belts consist of particles with energies extending into the relativistic regime. Relativistic particles have velocities near the speed of light and carry tremendous amounts of kinetic energy.

The ring current is so named because its charged particles produce an electric current that encircles Earth. Figure 4.3 is a schematic of the magnetosphere showing both the noon–midnight meridian and the equatorial plane. The solid arrows indicate the directions of the different currents flowing in the magnetosphere. Because of the shape and strength of Earth's dipole magnetic field region, energetic ions flow from midnight to the dusk side, and energetic electrons flow in the opposite direction. This difference in flow directions of positive charged ions and negative charged electrons gives rise to an electric current, a ring current that circles Earth. This ring current in turn gives rise to

---

[1] Amedeo Avogadro (1776–1856), Italian scientist and one of the founders of physical chemistry whose hypothesis was that a molar volume (a volume of gas whose mass is 1 gram molecular weight) of any gas contains the same number of atoms or molecules. This number, now called Avogadro's number, is equal to 6.022045×10^{23}.

[2] James Alfred Van Allen (1914–2006), American physicist and pioneer in the early development of the space program whose instrument on Explorer I (the first US satellite) discovered that Earth is surrounded by belts of trapped ionized high-energy particles. These regions are now called the Van Allen radiation belts.

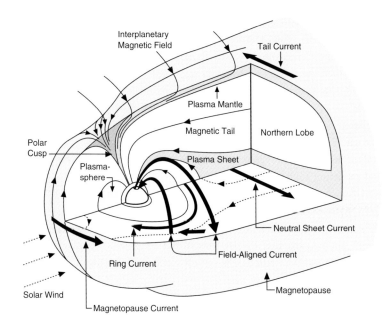

Figure 4.3 A schematic of Earth's magnetosphere showing the equatorial and noon–midnight meridional planes. The electric currents flowing in the magnetosphere are shown as dark arrows. The regions of the magnetosphere are labeled. (Adapted from Figure 1.18 in Kivelson and Russell, 1995)

a magnetic field that points in the opposite direction to the dipole field at Earth's surface. Therefore the ring current decreases the strength of Earth's magnetic field as measured on the surface. We use instruments called magnetometers near the equator to constantly measure the strength of the magnetic field. When the ring current intensifies suddenly, we see a rapid decrease in magnetic field strength. A magnetic index, called the Disturbed Storm Time index (or Dst), measures the deviation or change in Earth's magnetic field from its normal quiet time value, the strength of Earth's internal magnetic field. When this index goes negative (indicating a decrease in Earth's field), it is due to intensification or increase in the strength of the ring current.

Note that in Figure 4.3 there are other currents, called field-aligned currents, that connect the ring current and plasma sheet to the ionosphere. These currents, also called Birkeland[3] currents, play a major role in aurorae and other space weather phenomena and couple energy between the magnetosphere and the upper atmosphere of the Earth.

The radiation belts, named after their discoverer James Van Allen, consist of two distinct regions of energetic particles. The outer belt, composed mostly of energetic electrons, has its inner edge around 3

---

[3] Kristian Olaf Bernhard Birkeland (1867–1917), Norwegian physicist who made observations and developed the theory connecting upper atmospheric currents to the aurorae or northern lights.

Earth radii (r_E) and its highly variable outer edge usually just beyond geosynchronous orbit. The inner belt, which consists of energetic electrons and protons, extends out to about 2.5 r_E. The region between the belts (called the "slot") is generally kept clear of energetic particles by mechanisms that enhance the loss of the particles into the ionosphere. Plate 7 is a three-dimensional schematic of the donut- or torus-shaped radiation belts. The radiation belts contain intense radiation that can kill astronauts and damage or destroy sensitive electronics on spacecraft. Our understanding of this region has increased significantly with observations and modeling supported by NASA's Van Allen Storm Probes that operated from 2012 to 2019. Some of the most important insights learned from the mission included the role of different types of electromagnetic and plasma waves that create and intensify the belts and contribute significantly to the loss of electrons in the belts, hence explaining the dynamics of the Van Allen belts during space weather storms.

4.6 Interaction of the Solar Wind and Magnetosphere

Within the magnetosphere the dynamics of charged particles (the plasma) is determined by the configuration of Earth's magnetic field, which looks less and less like a dipole farther from Earth because of its interaction with the magnetized solar wind. The interaction of Earth's magnetic field with the magnetized solar wind is similar to that of a rock in a stream. The solar wind (stream) encounters Earth's magnetosphere (rock) as an obstacle and moves around it, leaving a wake behind. In the case of Earth and the solar wind the interaction produces a long magnetotail. Figure 4.4 is a schematic of Earth's magnetosphere.

Because the solar wind is supersonic, a shock wave is formed upstream or on the dayside of the magnetosphere. This shock wave is called the bow shock. The bow shock slows the solar wind and begins to divert it around the magnetosphere. The region between the bow shock and the magnetosphere is called the magnetosheath. The magnetopause is the boundary of the magnetosphere. The position of this boundary depends on the strength of the solar wind pressure, which is primarily due to solar wind density and velocity. As solar wind pressure increases, it moves the magnetopause Earthward. When solar wind pressure decreases, the entire magnetosphere expands. The location of the magnetopause is determined by a balance between the solar wind pressure and the magnetic pressure of the magnetosphere. The Earth's magnetosphere is analogous to the Sun's heliosphere.

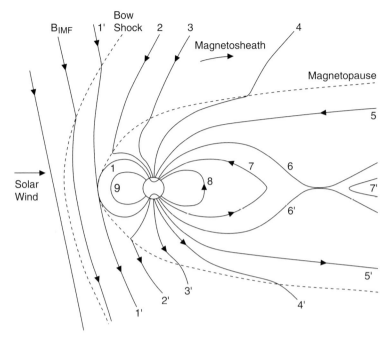

Figure 4.4 A two-dimensional noon–midnight meridional cross-section of Earth's magnetosphere showing magnetic reconnection on the dayside and in the magnetotail. (Adapted from Figure 9.11 in Kivelson and Russell, 1995)

4.7 Magnetic Reconnection

The above discussion treats the solar wind like a fluid (water in a stream moving around an obstacle). However, in the case of the solar wind, the fluid is magnetized, which makes interaction between the solar wind flow and Earth's magnetosphere a little more interesting. When two magnetic fields are brought together, the fields combine. So if you bring two magnets close to each other and measure the strength of the field at some point, you would measure contributions from both. However, magnetic fields have direction as well as magnitude, and therefore if a magnetic field points in one direction and another in the opposite direction, the two fields subtract from each other. When this occurs in a magnetized plasma, such as in the solar wind and magnetosphere, the fields can interact in a new way – to form new field lines. In this process, called **magnetic reconnection**, energy is taken from the magnetic field and put into particle motion (magnetic energy is converted to particle kinetic energy). This process, which has been reproduced inside the laboratory, powers much of the activity observed on the surface of the Sun. The NASA Magnetospheric Multi-Scale (MMS) Mission launched in 2015 has made distributed observations at high time resolution of magnetic reconnection at the magnetopause and in the Earth's plasma sheet, greatly improving our understanding of this process.

Magnetic reconnection occurs when a magnetized plasma parcel (often called a flux tube) of one polarity is brought up against a flux tube of opposite polarity. When magnetic reconnection occurs, the field lines connect and change their topologies or connectedness. Figure 4.4 shows examples of how field topology changes on the dayside magnetosphere. This figure shows an idealized magnetopause with the southward magnetic field of the solar wind (line 1′) brought up against the northward-directed magnetic field of Earth (line 1). Note that there are two distinct field lines, one that has both ends in the solar wind and one field line connected to both poles of Earth. When they come together, they can reconnect, and in addition to converting some of the magnetic energy into particle kinetic energy, the two original field lines' topologies are converted into two new field line topologies. Field lines (lines 2 and 2′) still exist, but one end of 2 is connected to the north pole of Earth and other end is in the solar wind, and one end of 2′ is connected to the south pole of Earth and the other end in the solar wind. Field lines with both ends connected to Earth are called "closed," and those with one end connected to Earth and the other in the solar wind are called "open." Plasma can become "trapped" on closed field lines, and therefore densities can build up. The plasmasphere and radiation belts are found on closed field lines. Open field lines generally have much less plasma since the plasma can stream along the magnetic field away from Earth.

4.8 Magnetotail

Reconnection of the Sun's and Earth's magnetic fields creates open field lines with one end attached to Earth and the other end extending into interplanetary space. Because the solar wind end of this magnetic field moves away from the Sun with the rest of the solar wind, the field line gets swept back behind Earth. This is similar to a person seated in a moving convertible car whose hair is swept back in the direction of the wind. Earth's magnetic field is swept back (lines 3, 4, and 5 in Figure 4.4) into a long, cylindrically shaped region called the magnetotail. The magnetotail consists of two magnetic lobe regions, one tied to the north polar cap (represented by line 5) and the other connected to the south polar cap (line 5′). These lobes contain "open" field lines of oppositely directed magnetic field – with the north lobe magnetic field pointing toward Earth and the south lobe magnetic field pointing away from Earth. These two lobe regions are separated by the plasma sheet – a region of higher plasma density than the lobes. This region carries the current that separates the two lobe fields called the cross-tail neutral sheet. These regions are labeled in Figures 4.2 and 4.3.

4.9 Plasma Sheet Convection

The solar wind imparts an electric field across the magnetosphere that is directed from dawn to dusk in the plasma sheet (the same direction as the neutral sheet current shown in Figure 4.3). This electric field causes the plasma sheet flux tubes to move Earthward in a motion called convection that completes the cycle brought about by magnetic reconnection on the dayside. Figure 4.4 shows the complete convection cycle or motion of a flux tube throughout the magnetosphere starting when a solar wind flux tube (line 1′) is first reconnected on the dayside magnetosphere (line 1). Because one end of the flux tube is connected to the solar wind, it is swept back over the polar cap (lines 2, 3, and 4). As the flux tube reaches Earth's nightside, the field line becomes part of the magnetotail (line 5) and is convected towards the central plasma sheet (line 6). At that point, magnetic reconnection takes two oppositely directed field lines – one from the north (line 6) and one from the south (line 6′) – and makes two new field lines (lines 7′ and 7). Line 7 on the Earthward-side of the reconnection site has both of its ends attached to Earth, while line 7′ now has both ends connected to the solar wind. Note that flux tubes represented by lines 7 and 7′ have the same magnetic topologies as lines 1 and 1′ at the start of the process. This means that reconnection disconnects the magnetotail plasma sheet from the Earth and it is ejected downtail. Line 7 is then convected Earthward to the dayside magnetosphere, where it can then participate in the convection cycle again.

Figure 4.5 shows the foot of the field lines projected into the polar cap of Earth. The figure represents a view down onto the polar cap with the numbers representing the position of the foot of each of the numbered field lines shown in Figure 4.4. The "convection cells" represent the motion of the field lines from the dayside over the polar cap and then back to the dayside at lower latitude. This convection motion, which is observed in the ionosphere, is the main dynamics of the ionosphere in the polar cap. The ionosphere and its dynamics are discussed more fully in Chapter 5.

4.10 Dynamics of the Magnetosphere

Reconnection (and hence convection) in Earth's magnetosphere is not steady. Enhanced reconnection at the dayside brought about by strong southward interplanetary magnetic field (IMF) can lead to increased energy coupling and an increase in the amount of magnetic flux transferred to the nightside. Increased magnetic energy density and pressure in the magnetotail lead to thinning of the current sheet, which enables

Figure 4.5 The projection of the magnetic field lines shown in Figure 4.4 onto the ionosphere. The lines show the direction of plasma motion in the ionosphere due to magnetospheric convection. (Adapted from Figure 9.11 in Kivelson and Russell, 1995)

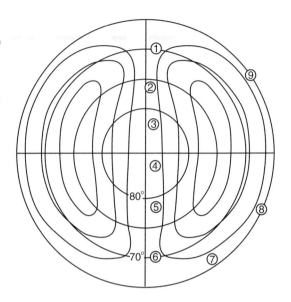

magnetic reconnection to occur. This in turn converts the magnetic energy of the tail into the plasma kinetic energy associated with rapid flows observed in the magnetotail. In addition, an enhanced cross-tail electric field leads to enhanced convection into the inner magnetosphere. This changes both the motion of particles and the location of the plasmapause; the plasmapause moves Earthward with increased convection, and away from Earth with decreased convection. Reconnection at the dayside requires a southward component to the IMF. Since IMF polarity points north and south irregularly, the amount of reconnection and hence energy input to the magnetosphere is highly variable.

In addition to the temporal variability, the magnetosphere and solar wind are highly spatially structured. Multiple satellite missions such as the European Space Agency's CLUSTER mission and NASA's THEMIS mission and new high-spatial-resolution global models have found that our simple pictures in Figures 4.2, 4.3, and 4.4 are highly smoothed. Keep this in mind as we discuss storms and substorms in the next subsections. The magnetosphere is highly structured both in time and space.

4.10.1 Storms

Occasionally, the amount of energy transferred into the magnetosphere from the Sun can increase rapidly. Such an increase is often associated

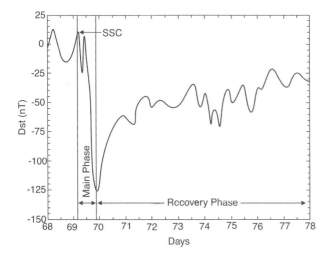

Figure 4.6 The Disturbed Storm Time (Dst) index over a ten-day interval showing the characteristic phases of a geomagnetic storm.

with the impact of Earth with southward IMF from a coronal mass ejection (CME). The southward IMF can reconnect with the northward field of Earth. This energy can increase the flow of energy and momentum into the magnetosphere and the rate of convection. Associated with this increased energy input is enhancement of the ring current. This is observed as a rapid decrease in the Dst. Figure 4.6 shows a time history of Dst during a storm.

A **geomagnetic storm** is often characterized by three phases: sudden storm commencement (SSC), main phase, and recovery. The SSC is characterized by an enhancement of Dst. This is due to an increase and motion Earthward of the magnetopause currents (named the Chapman–Ferraro[4] currents for the scientists who first hypothesized their existence and their role in geomagnetic storms [see Chapter 1]). As the CME and increased solar wind dynamic pressure hit Earth, the magnetopause moves Earthward. Chapman–Ferraro currents also increase to stand-off the solar wind by increasing the magnetosphere's magnetic pressure. The direction of the Chapman–Ferraro currents is such as to cause an increase in Earth's magnetic field as seen from the dayside low-latitude surface. Hence Dst increases. This enhancement generally lasts for tens of minutes to hours, when suddenly the rapid increase in the ring current swamps the Chapman–Ferraro current signal and Dst rapidly drops, signaling the main phase of the storm. The drop usually lasts several hours, at which time the Dst value begins a slow recovery to pre-storm

---

[4] Sydney Chapman (1888–1970), British geoscientist who made significant theoretical contributions to terrestrial and interplanetary magnetism, the ionosphere, and the aurora borealis. His ionospheric theory, called Chapman theory, explains the main structure of planetary ionospheres. Ferraro was a student of Chapman.

levels. The recovery phase can last for many days. The number of storms varies throughout the solar cycle, but typically is on the order of a few per month, with a great number and intensity of storms during solar maximum.

Associated with every storm is a rapid brightening and expansion of the entire auroral oval in both the northern and southern hemispheres. In many storms, the radiation belts intensify and the inner edge of the outer radiation belt moves Earthward. In some major storms, the slot region can be completely filled and the inner belt can intensify dramatically. The overall level of geomagnetic disturbance measured by another geomagnetic index called Kp also increases. Kp measures the overall variability of Earth's magnetic field observed at mid-latitudes. It is a logarithmic-scale (like the Richter Scale for earthquakes) and goes from 0 (no activity) to 9 (major storm activity). The average, or most probable, Kp level is about 3.

4.10.2 Substorms

Another smaller, but much more common, disturbance in Earth's magnetosphere is called a substorm. It is so named because it was originally thought that a collection of substorms makes a storm. Substorms occur much more often than storms – four times per day on average. Substorms are defined by auroral behavior. During a substorm, the most equator-ward existing auroral arc suddenly brightens and expands poleward and westward. The enhanced aurora is associated with enhanced auroral ionospheric currents that are measured by a magnetic index called the auroral electrojet (AE) index. The AE index is a measure of the difference between the strength of two ionospheric current systems, the westward and eastward electrojets. These electrojets, but especially the westward electrojet near midnight, are intensified by the onset of a substorm. An enhancement in the westward electrojet leads to a decrease in the horizontal field at Earth's surface directly underneath the aurora, while an enhancement of the eastward electrojet gives rise to an increase in Earth's magnetic field directly under the aurora.

Also associated with substorms are a number of other signatures including the presence of magnetic waves called Pi2 and the sudden enhancement of energetic particles at geosynchronous orbit. Figure 4.7 shows a schematic sequence of a substorm as seen in the global aurora. The panels show the polar cap from above, with midnight at the bottom of the panel, in the same format as in Figure 4.5. Note that the auroral disturbance begins near midnight and then expands poleward, eastward, and westward. The aurorae are the most visible manifestation of space weather, and the auroral display during a substorm is one of the most beautiful natural phenomena. Though they happen often, to see them

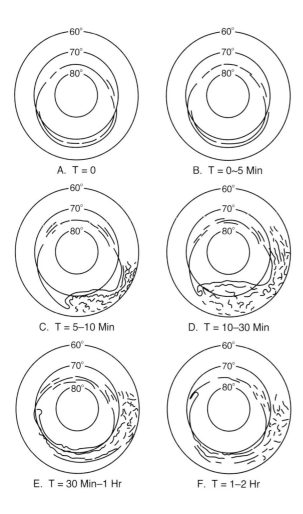

A. T = 0

B. T = 0~5 Min

C. T = 5–10 Min

D. T = 10–30 Min

E. T = 30 Min–1 Hr

F. T = 1–2 Hr

Figure 4.7 Configuration changes in the aurora, viewed looking down onto the polar cap, during a substorm. Noon is at the top and midnight at the bottom. Note that activity is concentrated on the nightside. (Reprinted from Akasofu, 1964, with permission from Elsevier)

from the ground, a number of conditions need to be satisfied: the substorm needs to occur where you are located, at night, and with clear skies. Since the auroral oval is at high latitudes, this means that aurorae are observed primarily during the winter from arctic countries. Aurorae are discussed in more detail in Chapter 5.

There are three main differences between storms and substorms: (1) their timescales – storms occur much less frequently and last for days, while substorms are common, occurring many times per day and generally having timescales of an hour; (2) their spatial extent – storms have global manifestations throughout the magnetosphere, while substorms are generally more localized to the nightside of Earth; and (3) their magnetic signature – by definition, storms are accompanied by an intensification of the ring current, whereas substorms are not. What adds to the complication and interest of storms and substorms is that all storms are accompanied by substorms, but not all substorms are associated with storms. Current research is attempting to distinguish the relationship between these two phenomena. For space weather research, the geomagnetic storm is of prime importance, though at high latitudes substorm currents and their impacts on the density structure of the ionosphere lead to important space weather impacts on power grids, pipelines, and airline communication and navigation.

4.11 Supplements

4.11.1 Additional Learning Objectives

After actively reading these supplements, readers will be able to:

- describe the relationship between electric charge and electric and magnetic fields, and how those fields exert a force on electric charges;
- calculate plasma, magnetic, and ram pressure of the solar wind and magnetosphere and be able to use that information to calculate the position of the magnetopause as a function of solar wind conditions.

Electricity and magnetism, which are intimately related, represent the two components of light and all other forms of electromagnetic radiation. The relationship between them can be described by a set of four equations called Maxwell's equations. Maxwell's equations also describe the density and motion of charged particles (currents), which are important concepts that have been discussed in terms of defining the space environment and coupling between the different regions. For simplicity, electricity and magnetism (and charge density and currents) will be discussed independently.

4.11.2 Electrostatics and Charge Density

Electric charge is an intrinsic property of matter. The subatomic building blocks of atoms are protons, neutrons, and electrons. Protons have a discrete positive charge, electrons have an equal but opposite negative charge, and neutrons have no charge (and hence are neutral). Most matter that we are familiar with on Earth (such as the book you are holding, and even you) is made up of atoms and molecules that have equal numbers of electrons and protons and hence are neutral. Electrons and protons are attracted to each other by their respective electric charge. This is the electric force or Coulomb force[5] that describes the force between electrically charged objects.

---

[5] Charles Coulomb (1736–1806), French physicist who developed the theoretical under standing of electric charge. The SI unit of charge (coulomb, C) is named in his honor.

Opposite (positive and negative) charges attract, and like (negative–negative and positive–positive) charges repel. To follow a convention first proposed by Benjamin Franklin,[6] electrons carry negative charge and protons positive charge. The strength of the electric force is dependent on the net amount and sign (positive or negative) of charge and inversely proportional to the square of the distance between the charged objects

$$F_c = \frac{(kq_1q_2)}{r^2},$$

where k is a constant of nature, q is the amount of net charge, and r is the distance between the objects.

Example:

(a) What are the magnitude and direction of the electric force between an electron and proton in a hydrogen atom? ($k = 8.99 \times 10^9$ N m^2 C^{-2}, the charge on a proton is 1.60×10^{-19} C, and the charge on an electron is equal, but with a negative sign. The average distance between the electron and proton in a hydrogen atom is 0.530×10^{-10} m.)

(b) If an electron mass is 9.11×10^{-31} kg, what is the acceleration of the electron due to this force? (Recall Newton's law, which states that $F = ma$.)

Answer:

(a) $F_c = \dfrac{(kq_1q_2)}{r^2}$

$$= 8.99 \times 10^9 \frac{\text{N m}^2}{\text{C}^2} \frac{(1.60 \times 10^{-19} \text{ C}) \times (-1.60 \times 10^{-19} \text{ C})}{(0.530 \times 10^{-10} \text{ m})^2}$$

$$= -8.19 \times 10^{-8} \text{ N}$$

toward each other (opposites attract).

(b) $F = ma$

$$a = \frac{F}{m}$$

$$= \frac{-8.19 \times 10^{-8} \text{ N}}{9.11 \times 10^{-31} \text{ kg}}$$

$$= -8.99 \times 10^{20} \text{ m s}^{-2},$$

which is a huge acceleration.

---

[6] Benjamin Franklin (1706–1790), one of the founders of the United States and a scientist whose work included proving the electrical nature of lightning (with his famous – and dangerous – experiment flying a kite during a thunderstorm).

4.11.3 Magnetostatics and Currents

Magnetism has been known for millennia, since lodestones, which are naturally occurring magnets, were discovered in a region that is now part of western Turkey called Magnesia (hence the name magnet). Magnets have the property of attracting or repelling one another and being attracted to certain metals such as iron. As mentioned in Chapter 1, the first western description of a compass was written in the eleventh century, and the realization that Earth was like a magnet was discovered in the years just prior to 1600.

All magnets attract iron, but because magnets have two poles, they either attract or repel each other. As discussed earlier in this chapter, magnetic poles are named north (+) or south (−), and opposite poles attract, like poles repel. Unlike electric charges (which come in either + or − varieties), magnetic "charges" always come together in a pair. It is impossible to separate the north pole from the south pole (i.e., cutting a magnet in half will not give you a separate north and separate south pole, but will give you two smaller magnets each with both a north and south pole). Even at the atomic and subatomic level, magnetic particles are dipoles, containing both polarities. A dipole magnetic field has a strength that falls off faster with distance than an electric field, whereas an electric field falls off as an inverse square of distance $(E \propto 1/r^2)$. Dipole magnetic fields fall off as the inverse cube of the distance, that is, $B \propto 1/r^3$.

Experiments conducted by Oersted[7] in the early 1800s demonstrated that electric currents deflect a compass needle. This had immediate impact for space weather as now there was a physical understanding of what could cause geomagnetic storms and deflection of compass needles by aurorae. This connection between magnetism and currents is what makes electromagnets and generators possible.

We now understand that all magnetism has as its source electrical currents. Within magnets, the currents are due to the alignment of individual atoms into domains. If these domains are ordered (aligned in a coherent manner with all the north poles pointing in the same direction), then the material is magnetized and will have a magnetic field associated with it. Electromagnets can be made by looping wire in such a way that the current flows in loops in the same direction. These current loops give rise to a magnetic field that is similar to a refrigerator magnet – both have a north and south pole, and you cannot cut the loops into two to make a stand-alone north or south pole.

---

[7] Hans Christian Oersted (1777–1851), Danish physicist who in 1820 demonstrated the connection between electricity and magnetism. This discovery led André Ampere and Michael Faraday to their major theoretical understanding of electromagnetism.

A magnetic field is a force field, similar to an electric field or a gravitational field, which exerts a force on an object at a distance from its source. Electric fields exert forces on electrically charged objects and particles. Gravitational fields exert forces on things with mass, while magnetic fields exert forces on other magnets and moving electrically charged objects or particles. Note the caveat with regards to electrically charged objects – they must be moving (relative to the magnetic field). One property of a moving charge is that it gives rise to an electrical current. Hence magnetic fields exert a force on a current.

4.11.4 Single-Particle Motion in a Magnetic Field

A single charged particle moving in a magnetic field will feel a force due to that magnetic field. The force is not in the direction of the field, as with gravitational fields and electric fields, but in a direction that is perpendicular to both the direction of motion and the field. This is expressed mathematically as a vector cross-product

$$\vec{F} = q\vec{v} \times \vec{B}.$$

Note the vector notation (arrows) above the three parameters – force, velocity, and magnetic field (magnetic field is written as B since M is used for another magnetic quantity called magnetization and m is used for mass). The symbol q, which represents the charge on the particle, is a scalar that is either positive or negative depending on the sign of the charged particle. This cross-product states that the direction of the force exerted on a moving charged particle in a magnetic field is perpendicular (at right angles) to both the velocity and magnetic field, and the magnitude is directly proportional to the component of the velocity that is perpendicular to the magnetic field. In other words, the magnitude of the cross-product can be written $F = qvB \sin \theta$, where θ (Greek letter theta) is the angle between the velocity vector and the magnetic field vector. Note that if the velocity vector of the charged particle is along the magnetic field, the magnitude of the force is zero since $\sin(0) = 0$. The magnitude of the force is a maximum when the direction of the velocity of the charged particle is perpendicular to the magnetic field. The direction can easily be found using the "right-hand rule" (see Figure 4.8). This rule states that you point your right-hand index finger (one closest to thumb) in the direction of motion, point your middle finger in the direction of the magnetic field, and your thumb points in the direction of the force (for negative charges this is in the opposite direction).

Figure 4.8 The Lorentz force is described by a vector cross-product $\vec{F} = q\vec{v} \times \vec{B}$. The right-hand rule helps determine the direction of the resulting force.

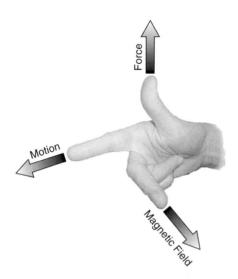

Note that the force will cause the particle to be accelerated (its velocity vector will change with time) because there is a force pushing it perpendicular to its original velocity direction. This causes the particle to circle or spiral around the magnetic field line. Particles with a velocity vector along the magnetic field line will not feel the magnetic force and travel along the magnetic field. Particles with a velocity vector perpendicular to the magnetic field will circle around the field line. Particles with part of their velocity vector perpendicular and part parallel to the field line will spiral around the magnetic field line in a helical trajectory.

Charged particles in a dipole magnetic field will have three types of motion – one due to the Lorentz force causing the particle to spiral around the field line (often called gyro or cyclotron motion), one due to the changing magnetic field strength (or pressure) along the field line that causes particles to move between the two hemispheres (called bounce motion), and one due to forces caused by the shape of the field line and the increasing change in magnetic pressure as a particle drifts closer to the Earth (called the gradient–curvature drift). These three types of motion are represented in Figure 4.9. Each motion has characteristic length and timescales associated with it. For example, the gyro or cyclotron motion has a characteristic gyro radius describing the size of the circular orbit around the field line, while the number of orbits it makes per second is called the gyro frequency (see Problem 4.5).

4.11.5 Pressure

Pressure balance between the solar wind pushing against the magnetosphere's internal pressure determines the size and shape of the

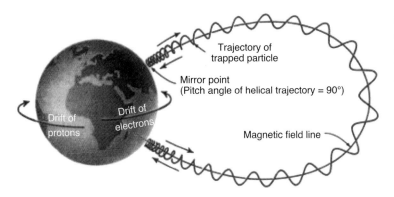

Figure 4.9 A charged particle undergoes three types of motion in a dipole magnetic field geometry: cyclotron or gyro motion around the field, bounce motion along the field, and drift motion around the Earth. (Source: Lawrence Livermore National Laboratory)

magnetosphere. Pressure is a force per unit area (units of newtons/m^2 or pascals in SI) and therefore describes how a force is distributed over a surface. A tire or balloon expands when the air inside exerts an outward force against the tire or balloon pushing against the external pressure. For a gas or fluid, the pressure is given by the ideal gas law ($p = nkT$, where p is pressure, n is the number of atoms or molecules per volume, k is Boltzmann's constant, and T is temperature). For a given volume, if you increase the density or temperature, the pressure will go up.

Similar to gas pressure, magnetic fields also exert a pressure that is proportional to the field strength squared

$$P_{\text{Mag}} = \left(\frac{B^2}{2\mu_0} \right),$$

where B is the magnetic field strength (tesla) and μ_0 is the permeability of free space ($\mu_0 = 4\pi \times 10^{-7}\,\text{H m}^{-1}$ [henry/m or newton/m^2]). Just like if you put more air in a balloon the balloon will expand, the size of the magnetosphere will expand if the magnetic field is stronger. For the case of the Earth, the strength of the main magnetic field does not change very much over space weather timescales, but the outside pressure from the solar wind does change quite a bit. Besides plasma and magnetic pressure, the solar wind also exerts a ram pressure due to the motion of the solar wind impinging against the Earth's magnetosphere. This pressure is similar to the force of wind on an object and is proportional to the mass density and the velocity squared:

$$P_{\text{Ram}} = \rho V^2.$$

This ram pressure actually dominates the plasma and magnetic pressure (over 99% of the total pressure). Therefore the solar wind ram pressure primarily determines the size of the magnetosphere. An equilibrium or balance is achieved when the ram pressure equals the magnetic pressure of the Earth's magnetosphere.

4.12 Problems

4.1 Which magnetic index is most often used to define when a geomagnetic storm occurs? What magnetospheric current does this index primarily measure? What behavior of this index indicates a storm?

4.2 Name three differences and three similarities between geomagnetic storms and substorms. What level of Bloom's Taxonomy does this question address?

4.3 What is the dipole magnetic field strength at the equator at geosynchronous orbit? (The equatorial field at the surface is 30 000 nT.)

4.4 Plasma is tied to magnetic flux tubes in magnetohydrodynamics. What is the direction of the electric field required for plasma in the plasmasphere to co-rotate with Earth? What is the direction of the electric field in the plasma sheet needed to convect plasma sunward? ($\vec{E} = -\vec{V} \times \vec{B}$)

4.5 L-shell is a magnetic coordinate system whose value is equal to the equatorial distance from the center of the Earth of a dipole field line. For example, L of 4 corresponds to the field line that extends out 4 r_E from the center of the Earth. What is the frequency of motion of a typical ring current ion at an L of 4? How does this change with distance from Earth? (The cyclotron frequency is $\omega_c = |q|B/m$, where q is the charge, B is the magnetic field strength, and m the mass of the ion.)

4.6 Estimate the location of the magnetopause at the subsolar point (a point directly between the Sun and Earth). Use the fact that Earth's magnetic field strength falls off as a dipole and that the magnetopause will be where the solar wind pressure (ρv^2) balances the magnetic field pressure ($B^2/2\mu_0$) of Earth's magnetosphere (where $\mu_0 = 4\pi \times 10^{-7}$ H m^{-1}).

4.7 Using the nominal solar wind and IMF values given in Table 3.1, calculate the total pressure of the solar wind (plasma, magnetic, and ram) to verify that the ram pressure dominates the solar wind pressure at 1 AU. (Hint: be sure to convert all parameters into SI units before calculating.)

(a) What is the ratio of plasma pressure to magnetic pressure on the surface of the Sun?

(b) What is the ratio of plasma pressure to magnetic pressure over a sunspot? (See Chapter 2 for typical photosphere parameters.)

4.8 Draw a concept map of the magnetosphere including its structure, regions, and dynamics.

Chapter 5
Earth's Upper Atmosphere

The region of the atmosphere in which this mirage effect originates has been called the Heaviside layer, an admirable title which is defective in only two particulars, first that the ascription to Heaviside is inexact and second that the region is not a layer. It might be permissible to call it the Balfour – Stewart – Fitzgerald – Heaviside – Kennelly – Zenneck – Schuster – Eccles – Larmor – Appleton space, but something less indirect is desirable. I have suggested the name ionosphere to make a systematic group troposphere, stratosphere, ionosphere, but meanwhile the term "upper conducting layers" seems to hold the field.

(Watt, 1929: the first use of the term ionosphere in the literature)

5.1 Key Concepts

- aurora
- ionosphere
- photoionization
- satellite radio communication and navigation

5.2 Learning Objectives

After actively reading this chapter, readers will be able to:

- describe mathematically the vertical structure, and the physical reasons for that structure, of the Earth's upper atmosphere;
- explain the reasons for the different layers of the Earth's ionosphere and describe how the ionosphere influences radio wave propagation;
- explain the mechanisms responsible for ionosphere variability and predict the geomagnetic conditions that give rise to the brightest and most dynamic aurorae.

5.3 Introduction

Earth's upper atmosphere plays an important role in ground-based and **satellite radio communication and navigation**, and its density determines the lifetime of low–Earth-orbiting (LEO) satellites. The upper atmosphere is composed of primarily neutral atoms and molecules in a region called the thermosphere. Within the thermosphere the amount of

83

ionized gas becomes appreciable and forms a region called the **iono-sphere** (see Figure 5.1). The thermosphere and ionosphere overlap in altitude, but because they describe two different particle populations (neutral and ionized) they are often "divided" since what influences the structure and motion of one usually does not necessarily directly drive the other. However, the two populations are coupled through particle collisions (neutral particle–ion interactions), which means that you usually cannot neglect one or the other. Since the thermosphere–ionosphere system is so important to radio wave propagation and LEO satellite lifetimes, it is one of the crucial areas of study for space weather.

5.4 The Thermosphere

5.4.1 Density Structure

Why does Earth have an atmosphere, while other planets, such as Mercury and Earth's moon, have essentially none? A strong force due to a pressure gradient tries to push any atmosphere near the surface of a planet or moon up and out into space. (Examples of this type of force are observed in a tire, balloon, or soda bottle. The pressure inside is higher than outside so there is a force wanting to push what is inside, air for the case of a balloon or tire, out.) Pressure is force per unit area; for a gas, it can be described by the familiar ideal gas law ($PV = nRT$ or $P = nkT$, where P is pressure, V is volume, n is the number of moles of gas, R is the gas constant, and T is temperature. In the alternate form of the same equation, n is the number of molecules per volume, and k is the Boltzmann constant). The ideal gas law states that the amount of pressure a gas will exert on its surroundings is proportional to the amount of gas and its temperature. A gradient is a change in a quantity as a function of distance. A pressure gradient means that the pressure in one place differs from that at another nearby place. This can be expressed as a simple algebraic difference relation: $P_1(x_1) - P_2(x_2)$, the difference in pressure at two different nearby points, or pressure gradient. The difference in some quantity over a distance is given a special mathematical symbol, called the del operator, written as ∇, so $-\nabla P =$ force. Note that the force of the pressure gradient points in the direction from high to low pressure (the reason for the negative sign is that the gradient points "uphill" always from low pressure to high pressure). On Earth the pressure near the surface is high compared to the relative vacuum of space so there is a force trying to move the air near the surface into space (the phrase "nature abhors a vacuum" [Spinoza, 1677] is because of the pressure gradient force). So why doesn't Earth's atmosphere get pushed out into space? The simple answer is gravity. Earth's mass exerts a force on the mass of the atmosphere attempting to pull it down to the surface. The balance of the upward pressure gradient force with the downward gravitational force determines the atmospheric density structure. This relationship is called hydrostatic equilibrium. The prefix "hydro" is Greek for "water" or fluid and "static" means "not changing." The word equilibrium means that the two forces exactly balance. We can write this relationship as

$$-\nabla P = \rho g,$$

where ρ is the mass density (units of mass per volume, or in SI, kg m^{-3}), and g is the gravitational acceleration (equal to 9.8 m s^{-2} at Earth's surface). Since pressure can be written in terms of density $P = nkT = (\rho kT/m)$ (since $\rho = nm$, where n is the number of molecules and m is their average mass), the hydrostatic equilibrium equation can be written in terms of the gradient in mass density (ρ) or number density (n). The solution to this equation is that the mass density and number density fall off as a function of height. The density falls off in a special way – exponentially. Density as a function of height can then be written $n(\text{height}) = n_0 \exp(-\text{height}/H)$, where H, called the scale height, depends on the composition (or make-up) of the gas (i.e., is it air, or pure oxygen, etc.), the temperature of the gas, and the acceleration of gravity, and n_0 is the density at the surface. The important thing to remember is that density falls off exponentially (quickly) with height. So with increasing altitude above Earth's surface, the amount of gas gets lower and lower.

Example: What fraction of sea-level air density is at the top of Mt. Everest? (Assume $H = 8$ km and height of Mt. Everest is 9 km.)

Answer:

$$n = n_0 \exp\left(-\frac{\text{height}}{H}\right)$$

$$\frac{n}{n_0} = \exp\left(-\frac{9}{8}\right) = 0.32,$$

or about 1/3 the density at sea level.

The Moon, Mercury, and Mars are much smaller than Earth, and thus the effective gravity at their surfaces is much less. Therefore the balance between the pressure gradient and gravity supports a much less dense atmosphere. Hence, if these solar system bodies had a thick atmosphere at one time, much of it could have escaped over their long history because the gravitational force isn't strong enough to hold the atmosphere to the planet or the Moon. The NASA MAVEN mission to

Mars launched in 2013 discovered that the atmosphere has been slowly stripped away through interactions with the solar wind. Since Mars does not have a magnetic field, it does not have an intrinsic magnetosphere to help protect the ionosphere from direct interaction with the solar wind.

5.4.2 Thermal Structure

When introducing the "spheres" in Chapter 1, we discussed briefly the temperature structure of Earth's atmosphere from the troposphere to the thermosphere and beyond. The thermosphere derives its name from the Greek word "thermos" (meaning "heat") due to its ability to absorb much of the high-energy electromagnetic radiation (UV and X-rays) from the Sun. This gives rise to temperatures ranging from 1000 to 2000 K depending on the solar cycle. As seen in Plate 3, during solar maximum the intensity of UV can increase by a factor of 10. The amount of X-ray emission can vary by over a factor of 100 between solar minimum and maximum. These large variations in solar high-energy radiation give rise to the factor of 2 temperature change of the thermosphere over the 11-year solar cycle. Since much of the UV and X-ray emission is from active regions and flares, there are also more rapid and transient heating events during solar maximum compared to solar minimum.

Besides UV and X-rays, the thermosphere can be heated by three other mechanisms: collisions with energetic charged particles, the flow of large-scale electrical currents in the ionosphere such as those associated with **aurorae** (in a process called joule heating), and energy dissipated through waves and tides that are set up by energy from the lower atmosphere and through interactions with the Moon–Sun system similar to ocean tides.

Energetic particles (cosmic rays, solar wind particles that can directly flow down the magnetic field in the cusp region, and particles associated with field-aligned currents in the auroral zone) can collide with thermospheric neutral particles transferring energy. Some of the energy can lead to ionization, excitation, and/or thermal motion. The increased energy or velocity provided to the neutral particle heats the thermosphere in a similar way to collisions with UV and X-rays.

Joule heating[1] is the process used in a toaster. When electrical current flows through a resistive wire some of the electrical energy is converted to heat, causing the wire to glow red hot. The higher the resistance (or lower the conductivity), the more energy is transformed to

---

[1] James Prescott Joule (1818–1889) was an English physicist who developed the concept of heat through mechanical motion that led to the idea of the conservation of energy. In addition to joule heating, the SI unit of energy (joule) is named in his honor.

heat. Electrical resistance describes how hard it is for electricity to flow. The resistance depends on the type of material the current is flowing through. For a toaster, the filaments are designed to have high resistance, which causes much of the electrical energy to be converted to heating the wire. As the ionospheric current flows, some of the charged particles that carry the current collide with the neutral thermospheric particles giving them energy through these collisions. Large-scale horizontal currents flow through the ionosphere mostly in the auroral zone associated with substorms, and hence joule heating is usually limited to the high-latitude thermosphere.

The final process of heating the thermosphere is from the dissipation of energy from waves and tides that contributes to enhanced collisions and turbulence. Waves called gravity waves can be launched from the troposphere upwards and they can break in the upper atmosphere dumping their energy in the thermosphere (similar to water waves breaking on the shore). Gravity waves[2] can be generated by convective thunderstorm activity and therefore are one mechanism that connects the troposphere to the upper atmosphere. The NASA ICON mission, launched in 2019, is designed to understand this energy coupling.

5.5 The Ionosphere

The upper atmosphere is usually defined as the region more than 80 km above the Earth's surface. At this altitude the density of neutral particles is low enough that free electrons, which are created through the process of ionization, can survive for an appreciable amount of time before recombining with ions. Ionization is the process of making positively or negatively charged atoms or molecules by adding or stripping away one or more electrons. In Earth's upper atmosphere it is much more common to make positively charged ions by removing an electron than it is to make negatively charged ions by adding an electron. Ionization is accomplished when electrons are knocked free of their host ion by either solar high-energy photons (mostly UV and X-rays) or energetic particles that precipitate into the atmosphere and collide with the surrounding gas. In the traditional model of the atom (called the Bohr[3]

---

[2] Gravity waves are small-scale buoyancy waves in the atmosphere and are not related to gravitational waves, which were first observed emanating from colliding black holes in 2016 by LIGO (Light Interferometer Gravitational-Wave Observatory) confirming one of Einstein's general relativity predictions.

[3] Niels Henrik David Bohr (1885–1962), Danish Nobel-Prize-winning physicist who developed the theory of atomic structure and explained the process of nuclear fission. The Bohr model of the atom describes the atom in terms of the nucleus (made of protons and neutrons) surrounded by orbiting electrons – analogous to a planetary solar system

model after the scientist who developed it), one or more electrons surround the nucleus, which is made of subatomic particles called protons and neutrons. Protons have positive charge, and electrons have negative charge of equal but opposite value. Opposite charges (positive and negative) have an attractive force called the electrostatic or Coulomb force, while the same charges (negative and negative or positive and positive) have a repulsive force. Almost all atoms and molecules in Earth's lower atmosphere are neutral, meaning that there are equal numbers of protons and electrons in each atom. In the upper atmosphere, the number of charged particles (ions and electrons) becomes appreciable. At altitudes of about 300 km there is a peak in the number of free ions and electrons. The region surrounding this peak in electron density is called the ionosphere. Figure 5.1 shows the vertical structure of this region.

The production of the main part of the ionosphere is primarily due to solar electromagnetic radiation through a process called **photoionization** and therefore the peak densities of the ionosphere are found on the dayside. However, at night the ionosphere does not completely go away since the recombination time of ions and electrons (the average amount of time needed for an ion and electron to come back together to form a neutral) is comparable to the rotation rate of Earth. The recombination rate is dependent on the background density, therefore the recombination rate is high at low altitudes (where the density is high) and decreases with altitude along with the density. The density of the ionosphere is primarily determined by the balance of the source or production of the ions (photoionization) with the loss (recombination) of the ions.

5.6 Ionospheric Structure

Photons of differing energies are able to penetrate and interact with atoms and molecules in Earth's atmosphere. Densities of atmospheric constituents (such as molecular nitrogen and hydrogen) also vary with height and therefore the ionosphere forms a number of different regions at different altitudes above the surface of Earth. Figure 5.1 shows how the ionosphere is divided into different layers or regions. Each region is characterized by a local maximum in the number density of ions. The D region, the lowest ionospheric layer in altitude, extends from approximately 50 km to 90 km (therefore it extends down into the mesosphere – see Figure 1.1). The main sources of ionization in the D region are solar

---

orbiting a star. Though quantum mechanics has changed our view of the atom, the Bohr model is still useful for understanding its basic structure.

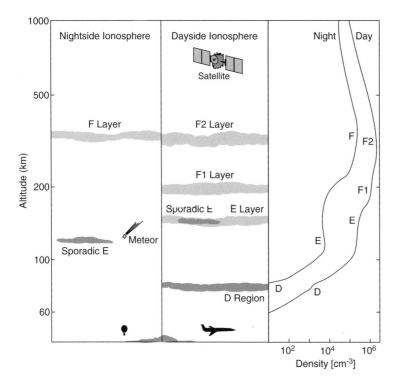

Figure 5.1 The vertical structure of Earth's ionosphere. Note that the ionosphere has several peaks labeled as D, E, and F layers. At night the D region normally essentially disappears, while the F region becomes a single layer. (Adapted from Radtel HF Radio Network)

UV photons ionizing nitric oxide (NO) molecules. During solar maximum conditions, solar hard X-rays ionize air molecules (molecular nitrogen and oxygen). In addition, cosmic rays produce ionization at this altitude. Because the neutral density is relatively high in the D region, the amount of recombination is very great. Therefore the D region is essentially only present during the day (though cosmic rays produce a residual level of ionization at night) and the level of ionization in the D region is the lowest of the different regions of the ionosphere. Solar storms can emit large amounts of X-rays that can cause rapid increases in D region ionization (sudden ionospheric disturbances [SIDs]). The D region is important with regard to high-frequency (HF) radio communication because it absorbs radio waves, which causes degradation of long-distance HF communication. During SIDs and intense polar cap precipitation of solar energetic particles, D region ionization can become so intense that HF radio communication is completely blacked out.

The next layer moving up in altitude is the E region (originally called the Kennelly–Heaviside or just the Heaviside layer). It extends from 90 km to 120 km and is formed by both low-energy (or soft) X-rays and UV solar radiation ionization of molecular oxygen (O_2).

The peak density in the E region is over 100 times greater than the peak density in the D region because recombination is less prevalent at these high altitudes. As with the D region, the E region decays away at night, which effectively raises its height as the faster recombination times at lower altitudes decay the E region away more quickly at low altitudes than at higher altitudes. Besides solar photons, E region ionization also occurs due to energetic particles precipitating into the atmosphere. Particle precipitation is particularly important at high latitudes. Impact ionization and excitation causes visible light to be emitted (the aurorae). The aurorae, which appear as ovals in the high-latitude northern and southern hemispheres, are some of the most beautiful natural color and light shows. Particle precipitation-induced ionization increases the E region ionosphere substantially, particularly at night when photon production is absent.

Other more transient sources of ionization at E region heights make thin regions of high-density ionization. These processes are lumped together to create what is called Sporadic E. Sporadic E are generally short-lived and spatially narrow regions of high ionization at E region altitudes. Sources range from complex dynamics due to the effects of the motion of the neutral atmosphere to meteors entering the upper atmosphere, which oblate (or burn up) and impact the surrounding neutral gas with enough energy to create an ionized trail. Sporadic E can last from a few minutes to several hours. Ionization can be locally very high, and therefore high-frequency radio waves can be reflected off these trails for long-distance communication.

The densest region of the ionosphere (and actually the entire magnetosphere) is the F region. It extends from 120 km and usually peaks at 300 km. The region above the peak is called the topside ionosphere, and the density decreases approximately exponentially with height and blends into the magnetospheric region called the plasmasphere. The transition between the topside ionosphere and the plasmasphere is typically at about 1000 km and is marked by the transition from oxygen as the dominant ion in the ionosphere to hydrogen as the dominant ion in the plasmasphere. (Plate 8 shows an image of the plasmasphere.) The F region is formed by extreme UV solar radiation ionizing atomic oxygen. F layer ionization decreases at night, but not as much as the E and D layer ionization because at this higher altitude recombination rates are lower and the layer consists of atomic oxygen rather than the molecular ions that dominate in the D and E regions. Atomic ions have much lower recombination rates in general than molecular ions.

Figure 5.1 shows schematically the vertical structure of the ionosphere. The middle panel shows the different layers that exist during the day. The F layer divides into two layers during the day because of the enhanced photoionization at high altitudes. The F2 peak is more dense than the F1 peak.

5.7 Ionospheric Variations

The ionosphere varies in systematic ways because the main source of ionization – solar UV and X-ray intensity – depends on the position of the Sun in the sky at a particular location on Earth and on the Sun's absolute output. When the Sun is directly overhead, the intensity of sunlight reaching the upper atmosphere is greatest. As the observer moves towards the poles or to the day–night terminator, the intensity decreases because the angle the Sun makes with the upper atmosphere is more oblique. As the observer moves into the dark or nightside hemisphere of Earth, the amount of sunlight goes to zero, and production due to photoionization ceases. The rotation and curvature of Earth therefore give rise to variations in the ionospheric structure. The largest cyclical variation is simply the day–night cycle, with dayside densities higher than nightside densities. There are also hemispheric (northern versus southern hemisphere) seasonal differences due to the continuous presence or absence of sunlight near the poles alternating between summer and winter.

In addition, the Sun's output of energy is not constant in time. It changes rapidly (especially at the high-energy end of the electromagnetic spectrum) due to solar flares and over the solar cycle. Plate 3 shows the Sun in UV emission over the solar cycle. Note that during solar minimum there is less UV emission, while at solar maximum the Sun's atmosphere emits large amounts of UV radiation. This gives rise to a solar cycle variation in the intensity of ionization of the ionosphere. During solar storms, the ionospheric structure can be drastically modified by energy input from the Sun. Therefore during geomagnetic storms the ionosphere becomes most disturbed and the most space weather impacts are noted.

5.8 The Aurora

As mentioned with regard to the E region ionosphere, energetic particles can precipitate into the atmosphere causing impact ionization and excitation that produces light. This light, called an aurora, is visible from the ground with the naked eye during the winter months in Earth's polar regions. Where do these particles come from and why do they primarily enter Earth at high latitudes? Recall from Chapter 4 that Earth's magnetic field is shaped like a dipole magnet. Field lines come out of the southern hemisphere and enter Earth in the northern hemisphere. Also recall that charged particles feel a force when moving through a magnetic field called the Lorentz force. This force causes particles to spiral around the field. If a particle is moving along a field line, there is no magnetic force exerted on it unless a component of its motion is perpendicular to or across the field line. Energetic particles trapped in Earth's magnetosphere are funneled down into the north and south poles where the field comes into and out of Earth. Therefore, impact ionization and creation of an aurora are most prevalent at high latitudes. The aurorae are also called the northern (and southern) lights, or aurora borealis (in the north) and aurora australis (in the south).

The aurorae appear as ovals around the poles (see Figures 1.2 and 4.7). This is because the particles that cause the aurorae are from the plasma sheet that occurs only in a narrow section of Earth's magnetosphere (see Figure 4.3). Aurorae can be several different colors (greens, purples, and reds dominate) because the aurorae are primarily from nitrogen and oxygen molecules and atoms. The color depends on what atom or molecule is excited by the impact of plasma sheet electrons and the energy of the impacting electron. The most common auroral color is green, which comes from excitation of atomic oxygen. Plate 9 shows the aurora as seen from Alaska. Figure 5.2 is a black and white image of the aurora over the upper Midwest and Northeastern USA. Individual bands or arcs as seen in Plate 9 make up the broader auroral oval, a part of which is seen in Figure 5.2. The aurorae are present constantly, but can undergo vast changes in brightness and extent. The most dynamic and widespread aurorae occur at night during substorms (Section 4.10.2).

5.9 Impacts on Communication

Recall from Chapter 1 that in 1901 Marconi sent the first transatlantic radio message. English physicist Heaviside and Irish physicist Kennelly proposed the existence of an ionized layer in the upper atmosphere to explain how Marconi's radio waves could be reflected around the curvature of Earth (the mirage effect mentioned by Watt in his paper proposing the term ionosphere in the epigraph of this chapter). The English physicist Appleton soon verified this suggestion experimentally. The ionosphere is still sometimes referred to as the Appleton layer.

How does the ionosphere interact with radio waves? Radio waves are electromagnetic energy with long wavelengths and low frequencies. As they propagate through an ionized medium, they become refracted

Figure 5.2 The visible aurora spreading from Minnesota to the Canadian Maritime provinces observed from 800 km altitude by a camera on the DMSP satellite. Note that cities on the east coast of the United States are clearly visible from space. (Image and data processing by NOAA's National Geophysical Data Center. DMSP data collected by the US Air Force Weather Agency)

or bent (Figure 5.3). The amplitude of the bending angle depends on the frequency of the electromagnetic wave and the density of the ionized gas. At a specific frequency – called the critical frequency – the wave will be perfectly reflected. This frequency, which is proportional to the density of the gas, is given by

$$f_{\text{critical}} = 9\sqrt{n_e},$$

where n_e is the electron number density in m^{-3} and the critical frequency is given in Hz. The peak ionospheric electron number density is typically 10^{12} m^{-3} and therefore the critical frequency is 9×10^6 Hz or 9 MHz. This is in the high-frequency (HF) radio band. At frequencies less than this critical frequency, the ionosphere will reflect a signal back to the ground (or if a signal is coming from space, back out into space). At frequencies above this one, the radio wave can propagate through the ionosphere (it will still be refracted, but won't be completely reflected). Since the peak density of the ionosphere changes with altitude, the critical frequency also changes. Therefore radio waves at different frequencies will reflect at different heights. This method of using

Figure 5.3 The ionosphere can refract and reflect radio waves. Long distance radio communication is possible due to "bouncing" radio waves off the ionosphere. (Adapted from Radtel HF Radio Network)

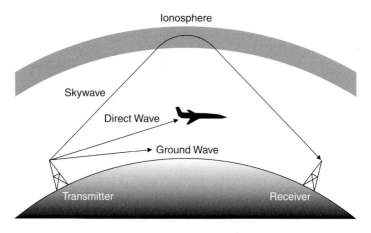

radio waves to sound the ionospheric density with height is still one of the primary observational methods to study the ionosphere.

5.10 Supplements

5.10.1 Additional Learning Objectives

After actively reading these supplements, readers will be able to:

- list and describe the different physical mechanisms from which atoms and molecules can gain energy through interactions with electromagnetic or particle interactions;
- calculate the density of the atmosphere as a function of altitude and explain the impact that changing atmospheric temperature would have on density at a given altitude.

5.10.2 Photochemistry

The main source of ionization and particle excitation in planetary atmospheres, including that of Earth, is photoionization. This is the process where a photon of electromagnetic radiation with sufficient energy (or frequency greater than a specific threshold) interacts with an atom or molecule and knocks it into a higher energy level. In general, photochemical reactions can be written schematically as $A + hf \rightarrow A^*$. This means that molecule A absorbs an amount of energy from an electromagnetic photon (represented by hf, the energy of a photon from $E = hf$, where h is the Planck constant and f is the frequency of the EM radiation) and undergoes a change of energy to an excited molecular energy state $A*$. This excited state could be energy transferred to the molecule's motion, called rotational or vibrational energy. Molecules can spin, or atoms in the molecule that are connected by chemical bonds can vibrate back and forth, analogous to atoms connected to each other via a spring. The molecule can also undergo an electronic transition, where an electron is knocked to a higher orbital shell, can be dissociated (broken apart), or can become ionized (have an electron knocked off or added). What is special about excited molecules is that their chemical reactivity is enhanced. In biologic systems this can have detrimental effects on the organism (see Chapter 8 for a further discussion of this), and in the atmosphere of Earth these excited molecules can change the local equilibrium state. One potential outcome of an excited molecule or atom is that it can transition back to the ground state by emitting a photon. This photon will have a discrete frequency depending on the atomic or molecular species and so this emission is "line emission." That is, the light from these transitions is in specific wavelengths or wavelength bands.

Atoms and molecules can also be excited by impact of electrons and ions. In this case, the optical emissions that result are termed aurorae since they are generally confined in geographic coordinates to the poles, where the magnetic field lines of Earth can funnel the particles from outside the atmosphere down into the ionosphere. The study of optical emissions from the atmosphere is called aeronomy (from the Greek for "the study of air"). Sydney Chapman coined this name, which now describes the field of space physics that studies the upper atmospheres of the planets where ionization and dissociation (the breaking apart of molecules) is important. Aeronomy is differentiated from meteorology – the study of tropospheric weather.

5.10.3 The Exponential and Hydrostatic Equilibrium

One of the most important mathematical functions in life and nature is the exponential. Exponential functions describe geometric series and the simplest dynamic systems. The function $f(x) = ab^x$ is an exponential function and consists of a constant b with the argument (x) as its exponent. The change in the value of the function (derivative) is directly related to the value of the function or in other words the growth rate depends on the value of the function. If "a" is 1, the value of $b =$ 2.718 ... = "e" and represents the natural exponential function. An example of a system that can be explained by an exponential is the density of the Earth's atmosphere (Section 5.4). The hydrostatic equilibrium equation can be solved in the following steps.

$$-\nabla P = \rho g$$

$$\frac{d\left(\dfrac{\rho k T}{m}\right)}{dz} = -\rho g$$

$$d\left(\frac{\rho k T}{m}\right) = -\rho g dz$$

$$-\frac{kT}{mg}\left(\frac{d\rho}{\rho}\right) = dz.$$

Defining $\dfrac{kT}{mg} = H$ and rearranging, we have

$$\frac{d\rho}{\rho} = -dz/H.$$

The solution to the derivative of a function divided by the function is the natural logarithm so

$$\ln\frac{\rho}{\rho_0} = \frac{-z}{H}$$

$$\frac{\rho}{\rho_0} = e^{-z/H}$$

$$\rho = \rho_0 e^{-z/H}.$$

Note that the change in density over a given distance depends on the value of density at the reference point ρ_0. This means that the magnitude of the change directly depends on the value of the reference density (if you start with a large reference density, the change in density over a scale height will be large). The exponential function describes not only density, but also population and the value of investments. There is an old saying that "making a million dollars is easy – after you have made the first $10 million" – it is the first $10 million that is hard. This is directly related to exponential functions. If I have $100 and earn 10% interest per year, after one year I'll have $110. If I have $10 000 000 dollars and earn the same 10% interest per year, after one year I have $11 000 000. The compound interest formula is a famous exponential equation and is

$$A = P\left(1 + \frac{r}{n}\right)^{nt},$$

where A is the final amount, P is the initial amount (principal), r is the growth rate, n is the number of time intervals that the change is happening over, and t is the total amount of time. To calculate an investment, A is the final amount of money, P is the amount initially invested, r is the interest rate earned, n is the number of compoundings per year, and t is the amount of time in years. Note that if $n = 1$ year (annual compounding), then the equation becomes

$$A = P(1 + r)^t.$$

So if your interest rate is 10% and your principal is $100, after 1 year your total amount of investment is $100(1 + 0.1) = \$100 + \$10 = \$110$. If you let the investment grow for 3 years, $A = \$100(1 + 0.1)^3 = \$100(1.1)^3 = \$133.10$ and if you let it sit for 7 years the value is $194.87, essentially double what you started with.

5.11 Problems

5.1 The amount of photoionization in Earth's ionosphere depends on a number of factors. Neglecting transport (motion of an ionized gas from one place to another), what are the most important factors that determine the amount of photoionization in the ionosphere?

5.2 The temperature of the thermosphere can reach 2000 K. Why aren't astronauts cooked as they "walk" in space?

5.3 What is the density of air at 100 km if the atmosphere has a scale-height of 8 km?

5.4 Atmospheric drag is proportional to density. How does the magnitude of atmospheric drag change over a solar cycle at 100 km altitude?

5.5 What effect would a major solar flare, which emits a large amount of UV and X-ray radiation, have on the dayside ionosphere?

5.6 How would Earth's vertical density structure change if (a) the Earth was one half its size, (b) the Earth was two times its size, (c) the Earth's atmospheric temperature was higher, (d) the Earth's atmospheric temperature was lower?

5.7 A satellite re-entering Earth's atmosphere will suffer a radio communications blackout because of the plasma created by the shock wave in front of it. If the satellite's radio operates at a frequency of 100 MHz, what is the minimum plasma density during the blackout?

5.8 If the number density of the Earth's atmosphere at the surface is 2.7×10^{25} molecules per cubic meter, what is the density at the top of Mount Everest (elevation about 8 km)? What is the density at 400 km altitude where the International Space Station orbits?

5.9 A dynamic system that can be explained by an exponential function with time (e.g., compound interest) can be related to a doubling time – the amount of time it takes to double the initial quantity. This is often called the "rule of 72." By dividing the interest rate by 72, you will find how long it takes to double. So a 7.2% annual growth rate would take 10 years to double. This works for population or investments or the number of operational satellites in orbit. The current (2019) annual growth rate for worldwide LEO satellite sales is about 20%. If there were approximately 100 satellites launched into LEO this year, when will we see over 800 launched? (How many doubling times needed?)

5.10 Draw a concept map of the Earth's upper atmosphere detailing its structure and dynamics.

Chapter 6
Technological Impacts of Space Storms

However, I would like to close by mentioning a possibility of the more remote future – perhaps half a century ahead. An "artificial satellite" at the correct distance from the earth would make one revolution every 24 hours; i.e., it would remain stationary above the same spot and would be within optical range of nearly half the earth's surface. Three repeater stations, 120 degrees apart in the correct orbit, could give television and microwave coverage to the entire planet. I'm afraid this isn't going to be of the slightest use to our post-war planners, but I think it is the ultimate solution to the problem.

(Arthur C. Clarke, Letter to the Editor, Wireless World, February 1945, p. 58)

This letter was the first detailed suggestion of global communication using geosynchronous orbit satellites. The prediction came true within 20 years with the launch of the first geosynchronous satellite in 1965.

6.1 Key Concepts

- atmospheric drag
- Faraday's law of induction
- radiation effects on satellites
- radio wave propagation

6.2 Learning Objectives

> After actively reading this chapter, readers will be able to:
>
> - list and describe the different types of satellite orbits around Earth and give examples of types of satellites that operate in each orbit;
> - describe and predict how satellite orbit lifetimes depend on their orbit type and orbital altitude;
> - list and describe the different types of radiation impacts on satellites and give examples of types of radiation hazards that are most prevalent in each type of orbit.

6.3 Introduction

Space weather has broad, everyday impacts on technology. Spacecraft and astronauts are directly exposed to intense radiation that can damage and disable systems and sicken or kill astronauts. Radio signals from satellites to ground communication and navigation systems, such as the Global Positioning System (GPS), are directly affected by changing space environment conditions. What may be surprising is that many ground systems such as power transmission grids, pipelines, and land-line communication networks (e.g., transoceanic fiber-optic cables) are also susceptible to space weather impacts. Plate 10 shows the wide variety of systems that are impacted by space weather. These include not only astronauts, but also commercial airline crew and passengers, as well as a host of satellite and radio communication effects. This chapter and Chapter 8 describe how space weather affects these systems and the impacts that space weather failures can have on technology and society.

6.4 Satellite Orbits

Our modern civilization has become dependent on space technology. We use satellites for a wide range of purposes including Earth observing (such as weather), communication (data, voice, television, and radio) and navigation (airplanes, ships, and driven and autonomous vehicles). You probably used a satellite today. You did if you connected to the internet through your smartphone, tracked a package being delivered to you by one of the major courier services, or used a credit card at a gas station pump or at a major retail store. Satellite technology is finding its way into a number of everyday activities and with 5G cellular communication, smart cities, the internet of things, and autonomous vehicles, satellites will play a larger and larger role in global connectivity. To support these services, there are hundreds of satellites orbiting Earth, and a number of communication companies have already launched hundreds more in the last few years with plans to launch thousands by 2025.

6.4.1 Types of Satellite Orbits

These satellites are in a variety of orbits, which means that each satellite has a unique path around Earth. Some satellites orbit close to Earth, others far from the surface. The orbit depends on the purpose of the satellite. There are four main classes of orbit for Earth-orbiting satellites, defined by the altitude above Earth that a satellite reaches. These are low Earth orbit (LEO), medium Earth orbit (MEO), high Earth orbit (HEO), and geosynchronous orbit (GEO). Figure 6.1 shows sample orbits for these four main classes. LEO satellites generally have circular orbits. A circular orbit means that the satellite's distance away from Earth's surface does not vary much during a complete orbit. Many

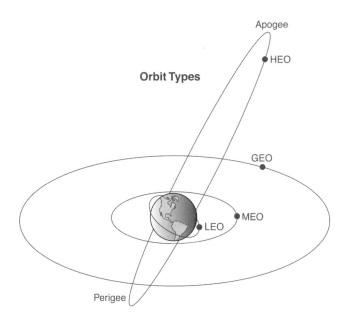

Orbit Types

Figure 6.1 The four main types of satellite orbits around Earth: low Earth orbit (LEO), medium Earth orbit (MEO), high Earth orbit (HEO), and geosynchronous orbit (GEO).

satellites have elliptical orbits, meaning that the satellite moves closer and farther away from Earth as it makes its way around Earth. In orbital dynamics, the distance farthest away from Earth that the satellite reaches is called apogee, while the closest is called perigee. The distance of the apogee determines the satellite's orbital period, the time it takes to complete one orbit. LEO satellites have an orbital period of about 90 minutes so circle the Earth about 16 times per day.

A medium Earth orbit (MEO) has its apogee above LEO and inside geosynchronous orbit (GEO). The Global Positioning Satellite (GPS) system has its satellite constellation in MEO. The GPS constellation allows multiple (at least four) satellites observable from any place on Earth enabling the triangulation of position. Because of the higher altitude compared to LEO, the orbital period is much longer than the typical 90-minute LEO and therefore enables multi-hour communication or observation passes to a single location from a single satellite. Medium Earth orbits can be circular or elliptical and have a variety of inclinations to the equator. Satellite inclination defines the angle that the satellite's orbit track makes with the geographic equator. High-inclination orbits spend time over the high-latitude regions of Earth, while low-inclination orbits stay near the equatorial region.

Geosynchronous satellites are at the right orbital altitude to have a 24-hour orbital period, the same orbital period as the Earth's rotation. Placed at the geographic equator they are geostationary and orbit above the same point on Earth continuously. They are primarily used as

Figure 6.2 Arthur C. Clarke's original sketch demonstrating the concept of global satellite communication using a constellation of three geosynchronous spacecraft. (Fig. 3 from Clarke 1945b, p. 306)

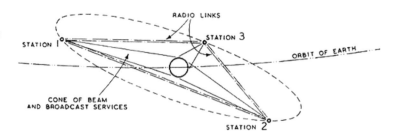

communication and weather satellites. By communicating between satellites, they can be used as part of a global relay communication system that enables nearly instantaneous communication around the Earth without needing to run communication cables. The path of the communication link runs from ground transmitter to satellite, then from satellite to satellite at other longitude, and then down to ground receiver. Figure 6.2 shows Arthur C. Clarke's original sketch showing how three geostationary satellites can be used to communicate globally.

High-Earth-orbit (HEO) satellites have their apogee above GEO satellites and can have a variety of ellipticity and orbital inclinations. The HEO shown in Figure 6.1 is highly elliptical with an inclined orbit that has its apogee high above the northern polar hemisphere. This Molniya orbit was designed for communication and remote sensing of the polar regions. GEO satellites observe the polar regions at low angles from their equatorial vantage; Molniya orbits provide long-time uninterrupted observational capabilities for high-latitude regions. Continuous coverage of the northern (or southern) polar region can be provided by two Molniya satellites with their orbits timed so that as one satellite approaches perigee in the opposite hemisphere, the other satellite is approaching apogee, enabling at least one satellite to be continuously above the polar region of interest.

6.4.2 LEO Satellite Purpose and Space Weather Impacts

LEO satellites have orbits from a few hundred kilometers to a few thousand kilometers altitude. There are a number of advantages of LEO satellites. The first is that a low Earth orbit is the easiest and least expensive orbit to get into. Depending on the size of the satellite, relatively small rockets can achieve LEO. Though they cost millions to tens of millions of dollars, these launch vehicles are a factor of 10 to 100 times cheaper than large rockets required to get some satellites into higher orbits. In recent years, a number of commercial launch companies have entered the market with some focusing on small payloads that have brought the cost down to under $1 million per small satellite launch. This has created new markets for other satellite-based services and is increasing the number of LEO satellites. Another advantage of LEO satellites is that the orbit is close to Earth, which means that small telescopes can see considerable detail on the surface, and low-power radio transmitters can easily send signals back to Earth. Therefore LEO is home to many Earth-observing satellites (such as those that provide pictures to Google-Earth®).

Atmospheric Drag

There are, however, several significant disadvantages to LEO. The main space weather disadvantage is that the lower the altitude, the more **atmospheric drag** there is on the spacecraft due to friction between the moving satellite and the upper atmosphere. Frictional force causes the spacecraft to lose altitude, which moves it into a denser neutral atmosphere (see Chapter 5 for the vertical structure of the atmosphere). The enhanced density causes increased drag, which lowers the satellite into even denser atmosphere and increases the orbital velocity. Eventually, the frictional force can heat up the satellite so much that it will begin to oblate or "burn up" in the atmosphere unless protected from the heat by strong insulating material. A satellite at less than 200 km altitude typically has a lifetime of only hours (one or two complete orbits). The International Space Station (ISS) normally has altitudes between 280 km and 460 km. The ISS requires propulsion systems and periodic reboost by resupply vehicles that allow it to raise orbit altitude regularly to prevent the space station from prematurely re-entering the atmosphere. Essentially all other LEO satellites do not have the capability to boost their orbits. Therefore all LEO satellites will eventually burn up in the atmosphere. With the retirement of the space shuttle in 2010, the Hubble[1] Space Telescope (HST) was no longer able to periodically dock with a shuttle to raise its orbit. It is expected that the HST will re-enter the Earth's atmosphere at the end of the 2020s or early 2030s.

The orbital lifetime of a LEO satellite depends primarily on its initial altitude and the density of the upper atmosphere (as well as the satellite's cross-sectional area). As discussed in Chapter 5, the density of the atmosphere is highly variable between solar minimum and solar maximum and therefore a satellite's lifetime also depends on when it

---

[1] Edwin Powell Hubble (1889–1953), American astronomer whose work established the existence of galaxies and that the universe is expanding. This was found by observing that galaxies are receding from Earth and that their recession velocities are directly proportional to their distance from Earth. This is called Hubble's law. The Hubble Space Telescope was named in his honor.

was launched. Space storms can cause rapid changes in the orbital altitude (and hence lifetime) of LEO satellites. The great storm of March 1989 caused thousands of space objects (including hundreds of operational satellites) to lose many kilometers of altitude. One satellite lost over 30 km of altitude (and hence a significant fraction of its orbital lifetime) during this storm. In February 2022, 40 of 49 Starlink communication satellites were lost immediately after launch when a geomagnetic storm caused an increase in thermospheric density causing them to re-enter the atmosphere before they could be boosted to higher altitudes.

The atmospheric orbital decay process has been extensively studied since it is one of the main parameters affecting satellite lifetime, and large satellites (such as the HST) in uncontrolled re-entry could crash into populated areas. Most satellites are small enough that they will completely burn up in the atmosphere and not reach the ground. However, pieces from large satellites (e.g., fragments from Skylab, which fell to Earth in July 1979) can survive re-entry and reach the ground. This would be similar to a large meteor entering Earth's atmosphere, and depending on the size of the fragment, could have tragic consequences if it hit an urban area. Large satellites are designed to have propulsion capabilities so that their re-entries are controlled. Many satellites do enter Earth's atmosphere, and pieces of them reach the surface, but by careful maneuvering at the end of the satellite's lifetime these are dumped into the ocean away from population centers. One such vehicle that was completed in 2010 and will eventually be decommissioned is the International Space Station (ISS). NASA has calculated that, during solar maximum conditions, the ISS loses 400 m of altitude per day (147 km per year). During solar minimum the loss is only 80 m per day (28 km per year). Therefore without periodic visits by resupply vehicles for fuel and reboosts, the ISS will relatively quickly re-enter Earth's atmosphere. Because it is so big, large pieces of the ISS will survive re-entry and hit the surface. Therefore, careful monitoring and control of the ISS is needed during its final orbits to make sure the pieces land harmlessly away from population centers.

Orbital Debris

There are tens of thousands of objects greater than 10 cm diameter in orbit around Earth. Only a small fraction of them are operational satellites; most of the debris is detritus of a space civilization (spent boosters, inoperable satellites, fragments) that uses satellites extensively for Earth observing, space science, communication, navigation, and national defense. Figure 6.3 shows the location of objects that are

(a)

Figure 6.3 A plot of orbital objects larger than 10 cm (a) out to geosynchronous orbit and (b) zoomed in to low Earth orbit. Each dot represents an object, but at this scale a dot is much larger than actual size giving a false impression that satellites and orbital debris fill space. Though there are many thousands of objects shown in these figures the actual space between objects in this snapshot is large. (Source: ESA and NASA)

(b)

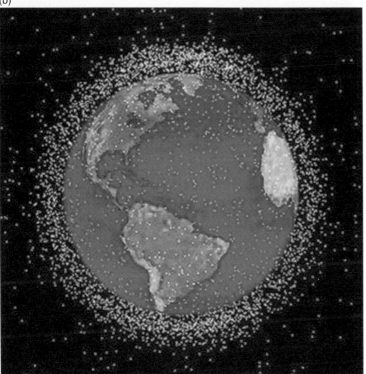

larger than 10 cm in orbit around Earth at a given snapshot in time: Figure 6.3a shows out to GEO, and Figure 6.3b zooms in to LEO. Note that at this scale it looks like space is completely filled, but in actuality close calls for collisions are considered when orbital paths cross within

5 × 25 × 5 km distance of one another (with the long distance along the orbit path of the satellites). If a prediction is made that the objects fall into this "alert box," further analysis is done and if the objects are predicted to be within a 2 × 5 × 2 km box then orbital collision avoidance maneuvers can be instituted. There have been several satellite–satellite collisions and several anti-satellite missile tests that destroyed satellites, creating large clouds of debris.

As part of international agreements, satellites in LEO are required to de-orbit within 25 years of the end of mission lifetime (though this duration is being reconsidered as orbital debris and collision likelihoods increase each year with more satellite launches). For many satellites this entails lowering the perigee of the spacecraft so that atmospheric drag increases, hastening the time when the satellite will break up into smaller pieces and then "burn up" (or more accurately ablate or vaporize) into the atmosphere. For satellites whose perigee is above LEO, their orbital lifetimes are essentially infinite and therefore objects will accumulate unless other orbital debris mitigation efforts are developed. For satellites in GEO, collision probabilities are not negligible, but one of the main drivers for dealing with satellites past their lifetime is that there are limited longitudes on which signals can be sent to specific markets (e.g., the continental USA). Because of the limits in the available frequency spectrum, only a small number of GEO satellites are able to broadcast to the entire continental USA. To make sure these orbit slots are not filled, an international treaty specifies that the last puffs of orbit-keeping propellant are used to boost the GEO satellite into a higher "graveyard" orbit so that a new satellite can be moved or launched into its place.

The main determination of low Earth orbital lifetime is thermospheric density at the satellite's orbital altitude. Density at a given altitude varies as a function of local time, orbital inclination, and solar and geomagnetic conditions. One of the goals of space weather is to understand how thermospheric density changes and its subsequent impact on satellite orbits and lifetimes.

6.5 Radiation Impacts on Satellites

Medium-Earth-orbit (MEO), high-Earth-orbit (HEO), and geosynchronous (GEO) satellites do not have significant satellite drag effects, but they have their own unique space weather concerns. These include spacecraft charging and high-energy radiation dose effects. Satellites in these orbits spend at least part of their orbit traversing the Van Allen radiation belts (discussed in Section 4.5), which contain trapped energetic particles that can severely damage or destroy sensitive electronic components. There are a wide variety of **radiation effects on satellites**. These include surface charging, deep dielectric charging, total ionizing dose, single-event upsets, and solar UV material degradation. The next subsections describe each of these effects in detail.

6.5.1 Surface Charging

Surface charging is due to the interaction between a spacecraft and the low-energy electron environment of space. A satellite in orbit will be impacted by both positively charged ions and negatively charged electrons. If the net transfer of positive or negative charge is not equal, net charging can take place. In addition to charged particle impacts, sunlight with enough energy can liberate electrons from any conducting surfaces via the photoelectric effect. The effect of these processes is that the spacecraft typically will become electrically charged. This is similar to scuffing your feet along a carpet and picking up electrical charge. If the spacecraft is made from or covered with different materials (such as the solar panel material and the main spacecraft body) then the spacecraft could have different parts charged up to different levels. If this happens an electric discharge (spark) may happen. This has a number of serious impacts. If the satellite is carrying any sensitive optical instruments, the spark could damage the detectors by overloading them with the bright flash. Another impact is that if the electric discharge is on a piece of sensitive electronics the component can be damaged or "fried." In order to prevent this from happening, care in the electrical design of the satellite is essential. Despite careful designs, a large number of satellites have experienced space weather-induced failures and effects due to charging.

6.5.2 Deep Dielectric Charging

Deep dielectric charging and discharging is one of the most common and catastrophic issues affecting spacecraft electronics and radiation. Relativistic electrons in the Van Allen belts have enough energy to penetrate the spacecraft and deposit their charge on the insulating material (or dielectric material) of the circuit boards that make up the electronic "brains" of the satellite. Electric charge can build up to such a level that the dielectric material breaks down, and charge can flow through these new pathways on the circuit board causing electrical shorts. Figure 6.4 shows a piece of plastic similar to that used in electronic circuit boards that has undergone dielectric breakdown. Had this contained electronic circuits, they would have been damaged. If the board plays a critical role

Figure 6.4 A sheet of plastic (like the material used in printed circuit boards in computers) after it was exposed to a large electric field. The dendril features are defects in the material due to the dielectric breakdown. (Used with permission from Bert Hickman, Stoneridge Engineering: www.teslamania.com)

in the operation of the satellite, the entire mission could be lost. Because of the potential catastrophic interaction of relativistic electrons with satellites, they are often called "killer electrons." A number of satellites have failed due to this space weather impact.

Satellite designers attempt to mitigate this effect by shielding sensitive components with thick aluminum covers or chassis and carefully grounding the circuit boards. However, the designers must attempt to design the spacecraft to be not only tolerant of penetrating radiation, but also as lightweight and small as possible since launches are so expensive. Therefore trade studies are done to find the optimal shielding thickness for the environment over the expected lifetime of the satellite. This is similar to designing a beach house in Florida. You can anticipate that the house may be subject to a category 3 hurricane in its lifetime, and so you design the house to withstand very strong winds. You don't build the house to withstand a category 5 hurricane because you don't want to live in a concrete bunker or pay for the extra heavy-duty construction required. Essentially you are conducting a trade study or risk assessment: what are the chances of various categories of storms? How much can I afford? How does hurricane-proofing the house impact its design? The architect will attempt to optimize the construction – design the best house possible with the resources available within the risk profile.

6.5.3 Total Ionizing Dose

Total ionizing dose (TID) is the cumulative exposure of a component or system to ionizing radiation (mostly electrons and protons). Electronic parts and devices can suffer degradation with radiation exposure over time that can result in failure. Some electronic devices are more sensitive to radiation damage than others and therefore several mitigation strategies are employed to enable the parts to work in the radiation environment that they are exposed to for their expected lifetime. This includes encapsulating the parts or enclosing the entire circuit board in aluminum housings of different thicknesses to provide shielding.

6.5.4 Single-Event Upsets

Single-event upsets (SEUs) are due to penetrating ions that can "trigger" an electronic circuit. Since ions carry charge, a number of detectors, switches, and current and voltage regulators can observe a pulse of charge as the particle interacts with the circuit. This can cause the switch or a computer memory bit to "flip," which could turn on or off or otherwise give an unintended signal to the spacecraft. These phantom commands can have catastrophic impacts. For example, a single-event upset in a logic circuit could provide a phantom command to fire the satellite's thrusters. Before ground controllers have figured out what happened, all the fuel could be spent, essentially ending the useful lifetime of the satellite. SEUs do not permanently damage the electronic part and are part of the class of single-event effects (SEEs) that include a single-event latch-up (SEL) and burnouts, which can permanently damage the component.

6.5.5 Solar Ultraviolet Material Degradation

Solar ultraviolet (UV) is much more intense in space than on the surface of Earth. The ozone layer and oxygen in our atmosphere are very effective absorbers of UV (and X-rays and gamma rays) and therefore the surface of Earth is shielded from much of this electromagnetic radiation. Life on Earth would be very different without this atmospheric shield since UV light damages living cells. In addition to the impact on living organisms, UV also can degrade certain materials particularly plastics and other organic materials. UV light is a contributor to solar cell degradation. In combination with energetic particle impacts (the main driver of solar cell degradation), UV light can make the solar panels less efficient. Satellite designers typically place solar arrays that are 25% bigger than needed for a particular mission, because over the lifetime of the satellite the efficiency of the arrays usually decreases by that amount. Individual solar storms can degrade solar cell efficiency by several percent and hence decrease the lifespan of a satellite by more than a year during just a single storm. Material selection for satellites, space stations, and future manned outposts on the Moon and Mars need to take into account increased UV exposure, which limits the types of material that can be used, especially organic polymers and plastics.

6.6 Radio Communication and Navigation Impacts

Space weather storms modify the density distribution of the ionosphere. Because **radio wave propagation** depends on the medium that the waves move through, a time variable and spatially inhomogeneous ionosphere can severely degrade and perturb ground-to-satellite and satellite-to-ground communication. This can have serious impacts for different systems, which are particularly important for high-frequency (HF) radio communication and Global Positioning System (GPS) navigation systems.

6.6.1 HF Radio Blackouts

HF radio is used for ship-to-shore and ship-to-ship communication as well as by commercial airlines for air-to-ground and ground-to-air communication. HF radio frequencies are between 3 and 30 MHz. This radio band is also popular with amateur radio operators. The ionosphere can reflect these frequencies and therefore long-range communication is possible by bouncing your signal off the ionosphere several hundred kilometers above Earth. This phenomenon, called "skywave," allows for over-the-horizon communication and is how Marconi was able to make the first transatlantic radio communication in 1901 (see Figure 5.3). The benefit of this frequency band – that it can interact with the ionosphere to permit long-range radio communication – is also its problem. Because the ionosphere is highly variable in space and time, HF radio communication can be severely degraded or even be made inoperable depending on a wide variety of factors. Many of these factors are related to space weather and include the amount of solar activity (and hence sunspot cycle) and geomagnetic activity (particularly aurorae).

  HF radio propagation depends on ionospheric density, which is controlled by sunlight and geomagnetic activity. Space weather degradation of HF radio has a particularly big impact on trans-polar airline flights. During large geomagnetic storms, HF radio communication can be rendered inoperable over the poles. Therefore commercial airlines, which rely on HF radio communication, must base their flight schedules on space weather forecasts. Airlines will reroute trans-polar flights during large geomagnetic storms because of the impact of the storms on their HF radio communication ability. Space weather not only affects HF communication systems, but can also affect satellite communication and satellite and magnetic navigation systems. Hence one of the primary space weather mitigation strategies for polar flights is simply to reroute to lower latitudes, although this causes longer flights and hence higher costs.

6.6.2 GNSS Satellite Errors

The global navigation satellite system (GNSS) consists of constellations of satellites operated by different countries to provide localization, navigation, and timing for military and civilian purposes. The US Global Positioning System (GPS), the Russian Global Navigation Satellite System (GLONASS) system, the Chinese BeiDou Navigational Satellite System (BDS), and the European Galileo system allow users to accurately locate their position anywhere on Earth. The GPS consists of over 28 satellites in medium Earth orbit arranged in such a way that at any given point on Earth at least four satellites are in view of an observer that has an unobstructed view of the sky. These satellites have atomic clocks on board and continuously broadcast the time. A user on the ground with a GPS receiver can receive this signal. By comparing the time broadcast by the satellite with the time at which it arrived, a distance (distance = speed of radio signal multiplied by the time for the signal to go from satellite to ground-user) to the satellite can be estimated. By triangulation (the process of determining the position of an object by using three independent distance determinations), the exact location of the user can be estimated. Because the user does not have an atomic clock, a fourth satellite is used to acquire accurate time, and three other satellites are used to triangulate position.

The speed at which a radio signal propagates through a vacuum is the speed of light (given as "c" in Einstein's famous equation). However, the speed at which an electromagnetic signal like a radio wave propagates through matter is less than the speed of light. This has the effect of slowing down and bending the signal, an effect called diffraction. The amount of bending and how much slowing occurs depend on the frequency of the signal and the properties of the medium. We experience this phenomenon when we look into water and when we see a rainbow. (Try this experiment: fill a glass with water and place a straw or pencil in the water. What happens to the straw or pencil when looked at from the side of the air–water boundary?) For a plasma, the property that determines electromagnetic propagation effects is the density. Therefore, because of the ionosphere, the radio signal from GPS is slowed down. GPS systems attempt to account for this delay by using estimates or models of ionospheric density. For typical handheld single-frequency GPS measurements, positional errors on the order of 10–50 m are common due to differences between the model ionosphere and the real ionosphere. This doesn't sound like much, but if GPS is used to fly an airplane, being 50 m off the runway can make a big difference. To correct for this effect, dual frequency GPS receivers can be used as well as differential GPS. The US Federal Aviation Administration (FAA) has developed a Wide-Area-Augmentation System (WAAS) that provides corrected location data to commercial airplanes to enable more accurate GPS positioning.

However, space weather can create small-scale ionospheric density irregularities that give rise to scintillation of the signal. Scintillation is the same phenomenon that causes stars to "twinkle." If scintillation is severe enough, the receiver can lose "lock" on the GPS signal, making accurate position information impossible. So the WAAS system is only as good as the GPS signal reception, and therefore severe space weather storms can make satellite navigation systems inaccurate and inoperable and can disable the entire North American WAAS system.

6.7 Ground System Impacts

A number of technological systems on the ground are susceptible to space weather. During a large geomagnetic storm, large time-varying currents flow into and through the ionosphere. These ionospheric currents can induce currents in the ground and long conductors on the surface. Electric power lines, telephone lines, and pipelines are examples of long conductors on the ground. Induced currents in these systems can overload electrical components, causing failure, or can decrease the lifetime of the infrastructure by enhancing corrosion. The main principle behind these induced currents is called Faraday's[2] law of induction. This physical relationship describes how a time-changing magnetic field can induce current and voltage in a conductor. Electricity can be described in terms of current or voltage, which are related through a relationship called Ohm's law. In space, electrical currents flow into, through, and out of the ionosphere. These currents intensify and move to lower latitudes during geomagnetic storms. The time-changing and spatially varying currents create a time-changing

---

[2] Michael Faraday (1791–1867), British physicist and chemist whose discovery of electromagnetic induction led to the invention of the electric generator and transformer.

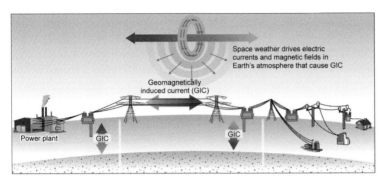

Figure 6.5 Currents flowing in the ionosphere can induce currents in the ground (geomagnetic induced currents, or GIC) that can flow through grounded electrical systems such as power lines. (Source: US Government Accountability Office)

magnetic field. According to Faraday's law, this time-changing magnetic field can then induce a voltage in long conductors and through the conductive layer of Earth's surface. These geomagnetic induced currents (GIC) can couple into ground electrical and pipeline systems. A power transmission wire is a good conductor designed to carry electrical signals long distances. On Earth, we have millions of kilometers of wire connecting buildings and houses with power plants and phone companies. These electric and communication grids are therefore susceptible to space weather effects. Figure 6.5 demonstrates how currents flowing in the ionosphere due to space weather events create GIC that can couple to power transmission systems and lead to damage in power transformers throughout the grid system.

6.7.1 Power Grids

Since the mid 1990s, power generation and distribution have become an interconnected continental-sized industry. Electricity produced by hydro-electric systems in Washington state in the United States is shipped to California. Power generated by HydroQuebec in eastern Canada can be shipped across the border to power homes in New York. Global power interconnections exist in South America, Africa, Europe, and Asia as well. Because of deregulation and this new interconnectivity, system vulnerabilities have increased. A power outage in one part of the grid can quickly propagate to other regions. Overgrown tree branches crossing a high-voltage line in Ohio triggered the power outage of 2003 that stretched from Detroit to New York City and left 50 million people in the dark. That same year, a solar storm likely caused a power outage in Sweden, which impacted the city of Malmoe. In March of 1989 a major geomagnetic storm caused an overload of a transformer in Quebec that quickly caused the collapse of the whole system. The transformer was exposed to ground-induced currents from the geomagnetic storm that exceeded its design capacity and it melted (see Plate 11). Transformers can convert high-voltage, low-current electricity into low-voltage, high-current electricity. It is more efficient to run high-voltage electricity long distances, but household appliances need high current. Therefore the electrical system ships the electricity from the power plant to the user at high voltage and then transformers located near the user convert the electricity into useful household or industrial high-current electricity. If the transformer gets more voltage than it is designed for (like from the induced voltage from enhanced ionospheric currents during a geomagnetic storm), it can fail. Power grid operators therefore must watch the geomagnetic or space weather forecasts and reduce the load on their systems during geomagnetic storms.

Of course if a storm occurs during a heat wave or cold snap when electricity usage is high, the operators may not have the flexibility to handle the situation and then must institute planned rolling "brown outs" or otherwise potentially suffer catastrophic blackouts. It is estimated that if a perfect storm occurs during the next solar maximum (a large geomagnetic storm occurring during a heavy electrical usage interval due to a cold snap or heat wave, or even during an interval of low electric use when more power is shuttled across larger distances), hundreds of transformers could be damaged or destroyed. Replacement could take years because transformer manufacturing is expensive and fairly limited. The US Department of Homeland Security estimates that such a storm could cost over $2 trillion dollars due to the severe economic disruption and could potentially cause civil unrest due to impacts on transportation, agriculture, retail, healthcare, water supplies, and communication systems.

6.7.2 Pipelines

Metal corrodes when exposed to a variety of environmental conditions (like moisture and air). Corrosion is enhanced if there is an electrical current flowing through the metal. A long pipeline can be susceptible to enhanced corrosion if electrical currents are allowed to flow across it.

Pipelines carry natural gas and oil throughout the arctic region from their source region to terminals at lower latitudes. For example, the Trans-Alaska Pipeline carries crude oil from Prudhoe Bay on the north slope of Alaska to the town of Valdez on the south coast of Alaska traversing a distance of nearly 1300 km (800 miles). In Valdez the oil is loaded onto super-tankers for shipment to California and refineries elsewhere. The pipeline sits underneath the auroral oval, which is coincident with the largest ionospheric currents usually seen due to geomagnetic activity. These time-changing ionospheric currents can induce large currents in the pipeline. The Alaskan pipeline is electrically grounded and protected to minimize this impact, but many pipes throughout the arctic region are not and therefore their lifetime and potential for leaks is increased because of space weather. Even for pipelines with electrical protection systems, large variations in GIC can cause the protection system to fail or not work properly. The major disruption of oil production in Prudhoe Bay in 2006 was due to severe pipeline corrosion that may have been exacerbated by currents induced by auroral activity and their impacts on the effectiveness of their protection systems.

6.8 Supplements

6.8.1 Additional Learning Objectives

After actively reading these supplements, readers will be able to:

- calculate the period and orbital velocity of a planet or satellite based on its apogee and describe why the orbital velocity is not dependent on the mass of the orbiting object;
- describe why LEO satellite orbits decay (lose altitude as a function of time) and explain how orbital velocity and atmospheric density change as a function of altitude;
- describe how an electric generator and electric motor work;
- identify components of a complex technological system and identify ways that they can interact and fail.

6.8.2 Kepler's Laws and Gravity

Johannes Kepler, using very precise data on the position of the planets in the night sky from Tycho Brahe, derived three laws of planetary motion that explained accurately the position of the planets about the Sun. Until Kepler, the positions of the planets were described by the Ptolemaic model of planetary motion. Because this was a geocentric model of the solar system (Earth was at the center), it required all types of geometric and mathematical tricks to describe the position of the planets. It was accepted because there was strong historical and religious motivation for having Earth (and hence mankind) at the center of the universe. Kepler's laws are based on the assumption that Earth orbits the Sun and not the other way around. The success of Kepler's laws was the death knell of the Earth-centered model of planetary motion and provided some of the strongest evidence up until that time

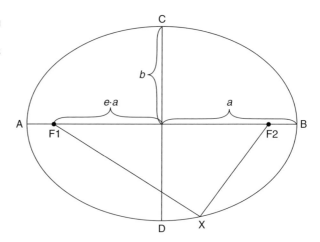

Figure 6.6 An ellipse showing the major axis (between points A and B) and minor axis (between points C and D). F1 and F2 are the foci. The semi-major axis is "*a*," while "*b*" is the semi-minor axis. The eccentricity is given by "*e*."

in support of the Copernican heliocentric model. The first of the three laws states that the planets travel in ellipses with the Sun at one of the foci. An ellipse is like a squashed circle – like an oval – and the two foci are located on the long axis that connects the edges farthest apart. This is called the major axis. The position of the foci depends on the length of the axis that connects the edges closest together. This is called the minor axis. The more squashed the ellipse, the farther the foci are from the center. (See Figure 6.6 for a diagram of an ellipse.)

The second law states that an imaginary line connecting the Sun and the planet around its elliptical orbit will sweep out equal areas in equal amounts of time. This means that as the planet moves around its orbit, it speeds up and slows down depending on whether it is at the point of its orbit that is closest to or farthest from the Sun. When the planet is closest to the Sun, it is at perihelion and it moves faster. When it is farthest from the Sun, it is at aphelion and moves slower.

The third law states that the period of the orbit (the time it takes the planet to go around the Sun once) is proportional to the distance the planet is away from the Sun. Actually the period squared is proportional to the distance cubed ($T^2 = ka^3$, where T is the period, a is the length of the semi-major axis, and k is a constant of proportionality). The constant of proportionality is equal to 1 if the units of period are given in years and the units of distance are in astronomical units (AU). For example, one can calculate the time it takes Jupiter (located 5 AU from the Sun) to make one orbit (one Jovian year) simply by solving $T^2 = a^3$ or $T = \sqrt{a^3} = \sqrt{5^3} = 11.2$ years.

Sir Isaac Newton developed the theory of gravity to explain the motion of objects falling on Earth and used it to explain the motion of the Moon about Earth and the planets around the Sun. The main concept of gravity is that two objects exert an attractive force (attempt to pull each other together) that is directly proportional to their mass and inversely proportional to the distance between the center of mass squared. Algebraically the magnitude of this force is written as

$$F_G = \left(\frac{Gm_1m_2}{r^2} \right),$$

where G is the universal gravitational constant, m_1 and m_2 are the respective masses of the two objects, and r is the distance between the centers of the masses. The direction of this force is always attractive. Most objects have their mass spread out through the object (i.e., the mass of Earth is spread out inside the volume of the spherical Earth). However, Newton showed that the mass of a distributed spherical body (such as Earth) can be thought of as a point mass (all the mass concentrated at a single point) at the center of mass of the object. The center of mass is the point that is the weighted average of the distribu-tion of mass for that object. For example, the center of mass of a uniform sphere, such as a baseball or Earth, to good approximation, is at the center. For a book, the center of mass is the point essentially halfway from each edge. For an object like a baseball bat, the center of mass is nearer the thicker part of the bat, but along the long axis. Therefore, the force of gravity felt by a person on Earth is directly proportional to the mass of Earth and the distance the person is away from the center of Earth.

If a mass is moving in a circle with constant speed (like a satellite orbiting the Earth) it has a changing velocity due to the direction of motion changing at each instant of time. The velocity vector is always tangential to the circular orbit path, similar to cyclotron motion of a charged particle in a magnetic field (Section 4.11.4). This change in velocity with time (or acceleration) is called centripetal acceleration. The centripetal force is

$$\vec{F} = m\vec{a} = \frac{mv^2}{\vec{r}}.$$

For a mass to stay in orbit, this force must be due to gravity, so

$$\left(\frac{Gm_1m_2}{r^2} \right) = \frac{mv^2}{\vec{r}}.$$

For a mass orbiting the Earth, $m_1 =$ mass of Earth and m_2 is canceled out since it appears on both sides of the equation. The velocity needed to stay in a circular orbit, as a function of distance away from the Earth, is therefore

$$v = \sqrt{\frac{GM}{r}}.$$

Note that G and M are constants and that this orbital velocity does not depend on the mass of the orbiting object. It depends inversely only on the distance from the center of mass. So as the orbital altitude increases, the orbital velocity decreases.

Newton's law of gravity confirms Kepler's laws and provides the physical understanding for planetary motion. Gravity is the force that determines the orbital characteristics of the planets (and artificial satellites), and the motion of the planets can be predicted to very high precision based on this relatively simple force law.

6.8.3 Atmospheric Drag

One of the main impacts of space weather on satellites in LEO is the impact of atmospheric drag on their orbital lifetime. The amount of satellite drag depends on their orbit altitude since the amount of atmospheric drag is proportional to the mass density of the atmosphere. Atmospheric drag acts as a friction (force is in the opposite direction to motion) and is felt when you are riding a bike. Have you ever tried to pedal faster and faster on a bike? As you increase speed, it becomes more and more difficult to speed up as the air drag is fighting against your pedaling. One way to reduce air drag is to bend over your handle bars to reduce your cross-sectional area. This is because air drag is dependent on four parameters including the cross-sectional area normal to the direction of motion (or relative direction of motion if the medium you are moving through is also moving e.g., if there is a wind blowing). The equation for air drag is

$$F(\text{air drag}) = AC_d\rho V^2,$$

where A is the cross-sectional area (m^2), C_d is the drag coefficient (a dimensionless number based on the shape of the object moving through the medium and ranging from 0.04 for the most aerodynamic shape to over 1 for cubes and other flat plate-like objects), ρ is the mass density (kg/m^3) of the medium, and V is the relative velocity of the object moving through the medium. Note that the drag force is non-linear with velocity – if you double your speed, the drag force increases by a factor of 4. So when biking, the amount of force needed to continue to accelerate increases non-linearly the faster you go. Also recall that the atmospheric density of the Earth falls off exponentially with height, or conversely increases exponentially as you decrease your altitude. Therefore a small difference in orbital altitude can have a large difference in the amount of air drag. This is compounded by a satellite's orbital velocity also increasing non-linearly as the altitude is decreased.

6.8.4 Current and Voltage in a Circuit

The modern world is completely dependent on ready and reliable access to electricity. Electricity powers almost every electrical appliance and system from cell phones, lights, computers, TVs, refrigerators, heating and air conditioning, traffic lights, to water pumps, etc. It is hard to imagine life without a cheap ready supply of electricity. Electricity is energy that can be converted into mechanical energy to run fans and compressors, or can power electric circuits that control microprocessors and radios. Most commercial electricity is generated by spinning large turbines (essentially fans) that convert the mechanical energy of the spinning fan blades into electricity through an effect called induction. Induction is the process that can make electricity when a magnetic field changes near a coil of wire and is the same process that gives rise to GIC. **Faraday's law of induction** mathematically describes this phenomenon. The fan blades in a turbine can be spinning magnets and the turbines are surrounded by many coils of wire. Usually the turbine's blades are spun by forcing steam through them, causing the fan blades to spin (essentially a steam engine). The steam is produced by boiling water, or by burning coal, natural gas, or oil. A nuclear plant uses the energy from splitting atoms to boil the water. Hydroelectric plants use rushing water instead of steam to spin the turbines (water wheels). Windmills use the flow of air past the fan blades to spin the turbine.

Since most power plants are large industrial sites, the power must be transported from the generating plant (usually located away from residential communities) to users. Electrical power is sent over long transmission lines as an alternating current (AC). Alternating current is electricity that oscillates in a sinusoidal fashion. In the United States the AC power oscillates at 60 Hz. AC is in contrast to direct current (DC), which is a steady state (not changing with time) or constant current. Batteries generate DC current. Electricity can be described by two main variables – current and voltage. They are related through the resistance of the material through which electricity is flowing. Resistance is essentially a measure of how easy (or hard) it is for electricity to flow. Metals and water have very low resistances (electric current flows easily), while wood, rubber, and air have high resistances (electric current has difficulty flowing). The relationship, called Ohm's[3] law, is that the voltage[4] equals the current times the resistance

$$V = IR.$$

One property of having electricity flow through a wire (like the electricity flowing the hundreds of kilometers from power plants to your house) is that part of the energy goes to heating the wire because of the resistance of the wire (joule heating and the same process that heats the thermosphere when ionospheric currents flow [Section 5.4]). The amount of power dissipated in a wire is dependent on the amount of current flowing in the wire. For a toaster, the filaments inside the toaster are high-resistance wire with large currents flowing in them. The current causes the wires to glow red-hot and toast your bread. For power transmission wires, we want to make the resistance as low as possible and make the current as low as possible. Therefore power plants send the electric power out along the transmission grid as high-voltage, low-current AC electricity. When it gets to your house, a device called a transformer converts it into a low-voltage, high-current source that is needed for appliances, computers, and lights. Because the power was transmitted with low current, the amount of energy loss due to heating of the transmission wires is minimized. Transformers use Faraday's law of induction (the same principle used to generate electricity) to change the voltage from high to low or low to high. The amount of voltage and current that a transformer can carry depends on its construction. Often transformers and power transmission grids operate near their peak capacity. If a large geomagnetic storm occurs during times of peak electrical usage, the system can become overloaded and the transformers destroyed.

6.8.5 Systems Engineering

Systems engineering is the technical approach that breaks down a technological system (like a satellite) into subsystems (e.g., communication, power, and sensors) and is the process used to design, build, test,

---

[3] Georg Simon Ohm (1789–1854), German physicist who deduced the theoretical explanation of electricity by careful experimentation and quantification of electrical current through a wire. The SI units of resistance (ohm) and conductivity (the inverse of resistivity called the mho – his name spelled backwards) are named in his honor (though the modern SI conductivity unit is called siemens (S)).

[4] Alessandro Giuseppe Antonio Anastasio Volta (1745–1827), Italian physicist credited with discovering how to make electricity and built the first battery. The SI unit of electric potential or electromotive force (volt) is named in his honor.

and operate technology. Similar to Systems Science (Section 1.6.5), the key is not only to understand the components or subsystems, but to understand how they interface and interact with each other. In a technological system, the design must fulfill the system's purpose (e.g., a communication satellite must enable communication between different ground and/or space systems) and meet all of the users' requirements. Requirements are technical specifications including size, weight, power, cost, accuracy, safety, etc. Requirements are essentially every parameter of the system that ensures the system works within its constraints. The constraints of the system could be the amount of money you have to spend (budget), the amount of energy or power that you can use, or the amount of risk (in terms of performance or schedule for example) that you can tolerate. Of course, there are often multiple designs or ways to achieve the same objective (both a mercury thermometer and a thermistor [electronic thermometer] can measure temperature), so how would you decide which is the best choice for your purpose? Systems engineering provides a systematic approach for conducting a trade study looking at many different options to meet the same performance requirements. The trade study looks at whether the solution meets all the requirements, how the choices impact all the subsystems and interfaces, and quantitatively assesses impacts on success, risk, and failure modes.

All technology fails. One aspect of good systems engineering design is to anticipate failure modes and attempt to eliminate, mitigate, and/or design graceful failure (e.g., fail-safe modes). One part of systems engineering is to understand all the critical components of each subsystem. The definition of a critical component is that if it fails, the entire subsystem and/or system fails. This could be a cable connector between the radio and the antenna or even the astronaut crew for a human mission to Mars. If the connector or astronauts fail (or are incapacitated or die) – the entire system or mission fails. These mission-critical components hence receive much more attention in the design, testing, and operation phase. Also note that there can be cascading failures when there are systems dependent on other systems that need to be clearly identified and considered. A cascading failure is when the failure of one system leads to the failure of other system(s). These can be present in any coupled or interconnected system including spacecraft and the human body. Space weather impacts can be catastrophic because technological systems that are susceptible to space storms include communication, navigation, and electrical power systems that are essential for modern society and are deeply interconnected to our transportation, energy, and economic systems.

6.9 Problems

6.1 Describe how space weather impacts airlines.

6.2 What are the orbital period and velocity of an astronaut in orbit at 300 km altitude? What is the orbital period of a satellite in geosynchronous orbit $(r = 6.6r_E)$? How long does it take the Moon to orbit Earth if it is at $60r_E$ from the center of Earth? (Use $k = 1.69$ and Kepler's third law with period (T) in hours and semi-major axis (a) in r_E.)

6.3 How does the critical frequency of Earth's ionosphere change from noon to midnight?

6.4 The type of orbit (LEO, MEO, HEO, or GEO) a satellite is in depends on its purpose. What are the advantages and disadvantages of LEO and GEO for communication and Earth-observing purposes?

6.5 Why are GIC most prevalent at high latitudes?

6.6 At a typical commercial airline altitude (10.6 km or 35 000 feet), how far can you see (or how far can line-of-sight radio communication travel)? Compare this to the size of an ocean or the polar regions.

6.7 What is the orbital velocity of a LEO satellite at 500 km altitude? What is the mass density of the thermosphere at this altitude assuming a scale height of 8 km (see Section 5.4.1)? What is the magnitude of atmospheric drag at this altitude in SI units?

6.8 What is the gravitational attraction of someone on Earth due to the Sun? How does this compare to the gravitational attraction of the Moon? Of the Earth? Of a classmate standing 0.1 m away?

6.9 Draw a concept map that organizes the different types of orbits and their space weather effects.

6.10 Identify a list of at least four parameters that are important for you with regards to transportation options for everyday commuting using the mode (walk, bike, bus, or car, or other options) around which to conduct a trade study. Identify strengths and weaknesses of each transport choice with respect to your other parameters and identify failure modes and their likelihood and consequence. Which parameters are most important for you in your decision?

Chapter 7
Space Weather Modeling and Forecasting

As the art of time-dependent global magnetospheric simulation develops, we expect it to be at first a source of new ideas ... and thereafter a source of quantitatively reliable information suitable for comparison with experiment.

(Leboeuf et al., 1978: the first global magnetohydrodynamic numerical simulation of the magnetosphere)

7.1 Key Concepts

- forecasting
- kinetic/MHD approach
- models
- simulations

7.2 Learning Objectives

After actively reading this chapter, readers will be able to:

- classify the different types of models, describe their assumptions, and give examples of space weather models in each type;
- create their own analytic and empirical space weather model using simple relationships and data sets;
- evaluate the strengths and weaknesses of different modeling approaches and specific models used in space weather research and forecasting.

7.3 Introduction

Since the 1980s, computer modeling of the Sun's atmosphere, the global interaction of the solar wind with the Earth's magnetosphere, and between the magnetosphere and the Earth's ionosphere and thermosphere, have progressed exponentially along with our observations, physical understanding, and computer technology. Conceptual and phenomenological models have been translated into physical models where assumptions and ideas can be tested and new experimental approaches developed.

The increased complexity and scale of space weather models now enables **forecasting** of solar and geomagnetic storms and their impact. The US National Space Weather Action Plan outlines the role of space weather forecasting and the development and use of advanced models to enable scientists to protect and safeguard critical infrastructure and lives that are susceptible to and threatened by space weather. The National Oceanic and Atmospheric Administration (NOAA) is home to the Space Weather Prediction Center (SWPC), analogous to NOAA's National Weather Service (NWS), which provides forecasts for space weather events to thousands of customers that are impacted by space weather. NASA's Coordinated Community Modeling Center (CCMC) provides resources to run, test, and share space weather models. This chapter describes modeling in general and some of the current state-of-the-art space weather modeling; the Problems section includes opportunities to create your own models, run a variety of models, and directly compare models with observations.

7.4 Models and Simulations

A model is a physical, mathematical, or logical representation of a system entity, phenomenon, or process. A simulation is the implementation of a model over time. A simulation brings a model to life and shows how a particular object or phenomenon will behave. It is useful for testing, analysis or training where real-world systems or concepts can be represented by a model.

(Lightsey, 2001, p. 178)

There is a hierarchy of **models** based on their complexity, ranging from toy models to fully coupled global physics-based models. Each type of model has its strengths and weaknesses, and each has a purpose or domain where its application is meaningful. Models enable scientists to explore extreme events and attempt to understand the key processes and mechanisms that are important.

7.4.1 Toy Models

Toy models usually attempt to replicate a system in simple ways (often analytical, or a logical or abstract representation). An example of a space weather toy model is the tipping-bucket model of geomagnetic substorms (see Figure 7.1 and Section 4.10.2). This model represents the loading and unloading of magnetic energy from the solar wind and IMF into the magnetosphere as water flowing into a bucket that is supported by a spring. Once the amount of energy (water) reaches

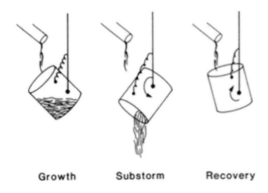

Figure 7.1 Toy models are often analytic models in that they use simple deterministic equations to describe the system. A tipping-bucket model has been used to explain the frequency of the occurrence of substorms.

Growth   Substorm   Recovery

a critical threshold, the bucket tips over and rapidly dumps the energy (water) from the system (bucket). This model can be represented by a few parameters – the energy (water) flow rate into the magnetosphere (bucket), the limit to the amount of energy the magnetosphere can store (size of the bucket and strength of the spring), or the amount of energy (water) that the system (bucket) can hold before becoming unstable (tipping over, emptying the water) and then returning to its initial state ready to take in more energy (water). A time-dependent model can then be described as the integral of the energy (water) into the system (bucket); it could be dependent on the north–south direction of the IMF, its magnitude and solar wind velocity and density, or what is often called a coupling function. Once the amount of energy stored in the system (bucket) exceeds a set threshold the bucket tips (an instability activates) and a geomagnetic substorm occurs. One can run this toy model over and over with different coupling functions, different instability thresholds and calculate a distribution of times between substorms based on solar wind and IMF drivers. This "simulated" distribution of times between substorms can be compared with the actual measured distribution to test our understanding of the system. It is a "toy" model as it clearly does not represent the entire complexity or physics of the system, but attempts to find a simple physical analogy that may provide insights. The use of an analogous physical system also helps in developing the important components of the system and potentially provides insight into its structure and processes.

7.4.2 Empirical Models

Statistical models using observations are called "empirical models" (from the Greek word for "based on experience"). They usually organize observations into meaningful "bins" representing physical locations, conditions, or time. For example, a spacecraft in elliptical orbit around Earth that makes measurements of the electron density can identify when it crosses the Earth's plasmapause (defined by a sharp radial density gradient with high density inside the plasmasphere and low density outside) usually twice per orbit (inbound and outbound) at different locations in local time, latitude, and geomagnetic conditions (see Section 4.5). Figure 7.2a shows the electron density observed by the CRRES satellite over one orbit with the location of the plasmapause (Lpp) being defined as the lowest L-shell location where the density drops by a factor of 5, at about Lpp = 4.7. (The L-shell is a magnetic coordinate system whose value is equal to the equatorial distance from the center of the Earth of a dipole field line.) For the entire mission, the Lpp was identified, and Figure 7.2b shows the locations as a function of geomagnetic activity, as indicated by geomagnetic index Kp (see Section 4.10.1). An empirical model was developed by fitting the data to a line and using the equation of the line as a simple analytic expression to best represent the fitted data. The slope of the line indicates that on average the location of the plasmapause (Lpp) decreases with increasing geomagnetic activity (as indicated by Kp). Note, however, the significant scatter in the data and that the best-fit line's location represents the average, but not necessarily an actual interval. Also notice that this model is dependent on only one variable (Kp) and hence assumes that other parameters (like local time or solar cycle) do not contribute significantly to the result.

Empirical models are extremely useful as they are often the best models that describe the system. Their weakness is that they are statistical, based on measurements that have intrinsic uncertainty, valid only under the conditions in which underlying data are sampled, and depend on assumptions that are made to create the analytic function. These introduce bias and uncertainties that ideally are recognized and stated. For example, one could make a model of how the strength of the Earth's magnetic field varies as a function of local time by taking large amounts of 1-second resolution magnetometer data from a single location, fitting a sinusoidal function to the data to determine the mean of each hour, and then fitting a quadratic expression to the data. Note that several choices have to be made in creating the model: what station should be used? Should every day of data be used or only selected intervals? Using all the data means including both days that are geomagnetic quiet (and hence representative of the Earth's internal field) and geomagnetic active days that include perturbations due to ionospheric and magneto-spheric currents such as the ring current. What criteria should be used in selecting which data to use? Is there a difference between making a sinusoidal, quadratic, or other mathematical function fit? Should

(a)

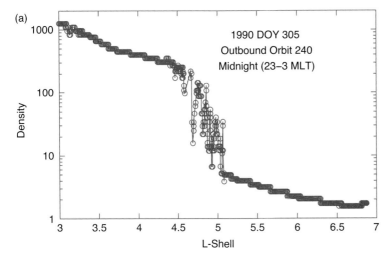

1990 DOY 305
Outbound Orbit 240
Midnight (23–3 MLT)

(b)

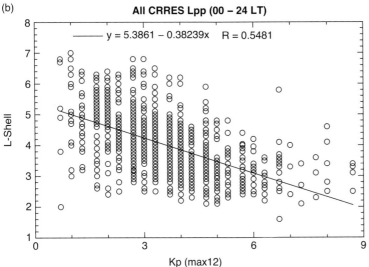

All CRRES Lpp (00 – 24 LT)

y = 5.3861 – 0.38239x   R = 0.5481

Figure 7.2 (a) An orbit of CRRES electron number density showing the location of the plasmapause as indicated by the sharp change in density over a short distance. (b) The location of the plasmapause observed over the entire CRRES mission organized by the level of geomagnetic activity (Kp) at the time of the plasmapause crossing. (Adapted from Moldwin et al., 2002)

1-second, 1-minute, or 1-hour averages be used? All of these choices have an impact on the resulting model.

Another issue of analytic models is that the data are often taken over only a limited range of parameter space, and often some variables are sampled much more than others. For example, moderate geomagnetic activity is much more prevalent than extreme geomagnetic storm intervals. Therefore, in a given time period (e.g., the 10-month lifetime of the CRRES mission) not all Kp levels were sampled equally. Orbital dynamics (or geographic location for ground observations) also introduces sampling bias. Spacecraft spend most of their time near apogee (see Section 6.4) and the local time (LT) of apogee precesses around the Earth with time (e.g., the LT of CRRES apogee started near 08 LT and over the lifetime of the orbit precessed through midnight to dusk to the noon sector). Therefore, the orbit itself introduces a sampling bias. Finally, the instrument will often have a certain range of parameter space that it can measure, limiting the amount of data taken depending on the scale of the variability of the parameter throughout its orbit or lifetime. For example, the CRRES plasma wave receiver used to measure the plasma density had an upper frequency limit that corresponds to a plasma density that was reached usually within an L-shell of 3 (see Section 5.9 for discussion on frequency and density). Therefore an analytic model of plasmasphere density would not have any data inside this distance. It is important to recognize the limits of validity (data or parameter space coverage) that went into making the model and not to extrapolate outside the domain of the model without caveats.

7.4.3 Physics-based Models

A physics-based model uses fundamental equations of physics to attempt to describe the behavior of the system. For example, a physics-based ionospheric model consists of dynamic (time-dependent) equations of density as a function of position (latitude, longitude, and altitude) and describes the forces acting on each parcel of plasma as a function of time. Newton's second law is expanded and can be written in a fluid approach to include all the important forces that contribute to the motion of the plasma. The continuity equation describes the creation (source), destruction (loss), and transport of plasma. Depending on the purpose of the model, these equations can be expanded to include more and more terms describing different physics and chemistry. For example, in the continuity equation there are only three ways of changing the density in a specific volume element as a function of time: move plasma into and/or out of the volume, create plasma, or destroy plasma. These terms are often called transport, source, and loss, and the continuity equation describes the only three ways to change anything physical with time. However, there can be multiple ways to create or destroy plasma (e.g., photoionization, impact ionization, recombination, and other chemical reactions) and hence there can be many equations needed for each of the three main terms.

$$\frac{\partial n}{\partial t} = \text{source} - \text{loss} + \text{transport}.$$

There are many advantages of physics-based models: they allow computer **simulation** experiments of conditions that are not possible to explore in laboratories (e.g., at extremely low densities of space or extremely high densities inside a planet) or answer questions about events of the past (how did the Earth's moon form?) or about events that we hope never take place (what would be the impact of a global nuclear war on ozone concentration?). In addition, they enable exploration of parameter space (what happens if we set this parameter to a small value or a large value?) or the important physics of a problem (what if we turn off gravity?).

7.5 Types of Space Weather Models

There are many types of space weather models, whether conceptual, toy, empirical, or physics-based. Within each of these classes of model there are a wide variety of models that attempt to explain specific regions, processes, or events. Sections 4.6–4.10 describe magnetospheric convection, which is a conceptual model that explains how the solar wind and IMF interact with the Earth's magnetosphere through magnetic reconnection. It can explain convective motion in the plasma sheet and polar ionosphere, and the occurrence of storms and substorms. This is called the "Dungey cycle" in honor of Jim Dungey, who first introduced this conceptual model. It makes testable predictions (e.g., enhanced convection requires southward IMF) and explains disparate observations (e.g., the two-cell convection pattern observed in the polar ionosphere, the disconnection of the magnetotail, and the creation of tailward-moving plasmoids – see Section 4.8, which describes how reconnection disconnects the magnetotail plasma sheet from the Earth and is ejected downtail. The disconnected plasma sheet could have a magnetic topology of a closed loop or flux rope that is called a plasmoid). This model has no equations or data – it is a conceptual model describing the cause and effect of reconnection in the magnetosphere. If I wanted to make quantitative predictions or understand the physical characteristics of the system (how much energy and magnetic flux is transported through the magnetosphere on what timescales), I would need to develop empirical or physics-based models based on data and theory.

Creating a global physics-based model of the entire system including the heliosphere, magnetosphere, ionosphere, and thermosphere is one of the penultimate goals of space weather modeling. Since the early 2000s, physics-based models have advanced using a variety of approaches: they are larger in terms of spatial scales (both in extent and resolution of the smallest scales) and faster in terms of the processes that can be resolved in the simulation and the computational time required to run a certain period of "real" time. A fluid approach (called magneto-hydrodynamics or the **MHD approach**) is the most developed as it focuses on the bulk large-scale motions and long timescales within the system and therefore requires the least computational resources. In another approach, particle codes track individual particles, or (in the case of particle-in-cell approach) super-particles that represent a collection of particles, throughout the simulation domain. These **kinetic** models require more computational resources as the number of particles needed to reproduce the distributions of plasma throughout a large spatial region is huge. Hybrid codes combine fluid and kinetic codes to attempt to incorporate all the important physics across a variety of scales.

One such coupled physics-based model that includes modules for different components and different types of approaches is called the Space Weather Modeling Framework (SWMF), which was developed at the University of Michigan. Figure 7.3 shows the different components of the model and how they are connected. The model uses a variety of approaches to represent a wide range of environments over a large range of scales and has been developed over decades by many investigators. Several other large-scale coupled space physics models have been developed at a number of institutions enabling inter-comparison of techniques, assumptions, and frameworks.

7.5.1 Magnetohydrodynamics

Space plasmas are a collection of different ionized elements (hydrogen, helium, oxygen, etc.) and equal numbers of electrons. Each of these particles can have its own velocity, charge, mass, and energy. Air in the atmosphere of Earth or in the room in which you are reading is made of neutral gases (molecular nitrogen and oxygen primarily), each of which has its own velocity. One way of determining the motion of this collection of particles is to follow each individual particle in its path around the room, keeping track of its acceleration, instantaneous velocity, and position. This is a monumental task if you consider the number of particles in any given volume of space. Fortunately, gases and plasmas often behave in a collective way, which is called the fluid approximation. Fluids move collectively – you can describe the motion of a river in terms of its velocity at a point instead of having to measure each and every molecule of water at that point. Because water molecules act collectively on scales that are large compared to their size, we can follow the motion with a set of simple

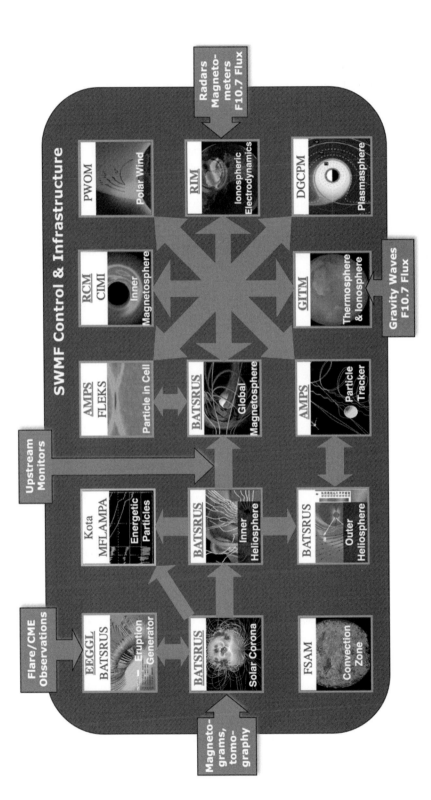

Figure 7.3 The components of the space weather modeling framework (SWMF) showing how the different models are coupled. This model is used by NOAA SWPC as the operational space weather forecast model to help predict and understand space weather storms.

equations that keep track of a volume of the fluid rather than the individual particles. The most important of these are the continuity equation and the momentum equation. The continuity equation (Section 7.4.3) states that if the density of a fluid changes with time in a given volume, it can be due to three different processes: fluid has moved in or out of the volume (transport), fluid was created in the volume (source), or fluid was destroyed in the volume (loss). The momentum equation describes the motion (or transport) of the plasma and how the velocity of a parcel of fluid changes with time. Since velocity can change with time (acceleration) because of an applied force, this equation essentially lists all the forces that act on the fluid parcel.

This treatment can be done for magnetized plasmas as well. But, unlike neutral fluids such as water in a stream, plasmas are made up of charged particles. Therefore, the forces felt by the plasmas include electric and magnetic forces. The simplest approximation for a plasma treats the entire collection of particles (ions and electrons) as a single fluid. Referred to as the magnetohydrodynamic (MHD) approximation, it describes the large-scale dynamics of plasmas (such as the motion of the solar wind throughout the heliosphere and the plasma inside Earth's magnetosphere) very well. The main forces that act on a parcel of plasma are the pressure gradient force discussed in Chapter 5 and the magnetic force discussed in Chapter 4.

7.5.2 Kinetics

Three main assumptions of MHD are that the spatial scales being modeled are large compared to length scales representative of ion motion (e.g., the ion gyro radius – Section 4.11.3), the timescales are long compared to ion motion (e.g., the ion gyro frequency), and the matter being modeled acts collectively as a fluid. If you are interested in modeling small-scale, fast and/ or individual particle motion you must use equations that describe the motion of the individual particles. These particle kinetic models can essentially use the same equations of MHD, but instead of tracking fluid elements, describe the motion of either single particles (Section 4.11.3) or collections of particles described by a distribution function that keeps track as a function of time of how many particles are located at a certain position with a certain velocity. As computational resources and capabilities continue to increase with time, larger spatial domains can be modeled without making the MHD assumptions to model both the small and fast scales on a regional or even global scale. However, currently many modeling approaches use "hybrid-models" that use MHD at the global scale and insert kinetic models in regions where there are sharp changes of parameters over small distances (sharp gradients) or in regions where electron motion is important.

7.6 Forecasting

Using models to understand how certain past and current conditions determine our future conditions is one of the hallmarks of science and one of the greatest achievements of human thought. Numerical forecasting, or scientifically predicting the future, has enabled understanding of our natural surrounding and our lives and has powered our technological developments around healthcare, sanitation, transportation, agriculture, communication, and housing. Being able to predict the weather not only enables you to know if you should bring an umbrella with you, but also saves lives by predicting severe weather.

Space weather forecasting is like weather forecasting in that it combines observations with a variety of models to attempt to characterize the current state of the system and predict its future state with a variety of lead times. Clearly the longer the lead time of an accurate forecast, the more options that the end user has in how to respond (e.g., an evacuation order for a hurricane is not useful if it is given only a few hours prior to landfall, while if it is made days before landfall many more people can be moved to safety). For all forecasts of complex systems there is an inherent limit on the accuracy of the forecast over multiple dimensions – spatial, temporal, and severity. Using the hurricane analogy again, to be useful a forecast for hurricane landfall needs to accurately predict "where" the hurricane will hit, "when" it will hit, and the magnitude of the impacts (storm surge, rainfall, winds, etc.). The farther in the future a forecast is made, the less likely it will be accurate. Therefore, many long-range forecasts predict general quantities and properties instead of specific details of an event. For example, there are groups that forecast the number of named storms that will happen in the Atlantic hurricane season. They do not attempt to forecast individual storms, just the number and perhaps range of severity (number of tropical storms vs major hurricanes).

For space weather, the same range or type of forecasts are made. For long- term forecasts, there are groups that predict the sunspot number as a function of time for the next solar cycle or estimate the probability of a major Carrington-class geomagnetic storm for any given year. There are also groups that attempt to take observations of the Sun, the solar wind, and interplanetary magnetic field to forecast the timing and severity of geomagnetic storms and even regional forecasts for different locations. These predictions can use physics-based models, empirical or statistical models, or even machine-learning models that combine observations with physics-based models.

7.7 Supplements

7.7.1 Additional Learning Objectives

After actively reading these supplements, readers will be able to:

- define machine learning and understand how improvements in computation and increased numbers of observations have led to new modeling techniques;
- assess the strengths and weaknesses of scientific (and other) arguments.

7.7.2 Moore's Law and Bell's Law

In 1965, the co-founder of Intel noticed that the number of transistors that fit on a computer chip appeared to double every two years (see Section 5.10.3 for discussion of exponentials and doubling times). Each transistor on an integrated circuit (IC) can perform an operation and therefore the more transistors, the more operations. Figure 7.4 shows the growth of transistors over time.

The exponential increase of computing power (and memory) has essentially held to the present through new innovations in materials, nanotechnology, new geometries, and now with the advent of quantum computing.

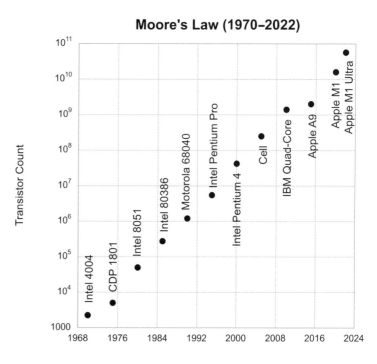

Figure 7.4 Moore's law states that the number of transistors, or operations that a computer chip can make, doubles approximately every two years. (Adapted from OurWorldinData.org, licensed under CC-BY-SA by the author Max Roser)

With the increase in computer speed and memory, larger models with fewer assumptions can be used. For example, global MHD models are able to simulate the entire magnetosphere, but assume slow processes over long length scales (fluid approximations). Kinetic models follow individual particles (or distributions) and therefore can model the small-scale short-timescale processes (e.g., reconnection) at boundaries (e.g., current sheets) and therefore provide more physically realistic models. However, because of the amount of computation needed, kinetic models often make approximations themselves (e.g., often the mass ratio between ions and electrons is set significantly smaller than in reality) and often are limited in spatial scale. As computing power grows, fully kinetic global modeling becomes an option. The next issue is how to manage, visualize, and interpret all the results!

Bell's law states that a new class of computer develops every decade and becomes smaller, cheaper, more energy efficient, and more ubiquitous. Named by Gordon Bell (electrical engineering at Digital Equipment Corporation (DEC)) in 1972, it highlights the evolution of computer architecture with time. In terms of modeling, this is opposite to Moore's law in that it predicts less and less computing power on a single device, but enables new classes of experiments, sensors, and distributed networks of observations. Each new class is observed (and predicted) to develop over a decade enabling new types of approaches and possibilities. The "internet of things" (IoT) is a world of distributed processors, controllers, and sensors that when interconnected enable real-time monitoring of houses, buildings, cities, and the environment. As we move into the mid twenty-first century, global cloud computing, distributed arrays of instruments, and constellations of satellites could combine to provide users with the ability to assess and predict the state of the space environment across scales. Data assimilation, machine-learning, and physics-based models running in real time could provide high-fidelity representations of the system state for space weather operations and space physics research. If Moore's law and Bell's law continue, what will the state of our connection to computers be in 10, 50, or 100 years?

7.7.3 Statistical Modeling, Machine Learning, and Artificial Intelligence

Empirical modeling uses observations to understand systems. For example, the location of the Earth's magnetopause or plasmapause has been modeled by examining how the radial location from the center of the Earth changes as a function of local time, solar wind conditions, and/or geomagnetic conditions. These models combine boundary observations from many different satellite orbits over time to build up a large number of observations to determine the distribution of observations at a given local time and conditions. If the distribution is a normal distribution, the average or mean +/– the standard deviation gives a statistical location. If the distribution is not normal, then other statistical characterizations can be used such as median and quartiles or deciles. Examining linear or non-linear relationships between solar wind drivers and geomagnetic response has been done extensively to develop "coupling functions" that show which solar wind drivers are most responsible for geomagnetic response. The Akasofu epsilon parameter was one of the early derivations of a coupling function. This approach requires the investigator to explore a variety of parameters and mathematical expressions to fit the data. What if we could let the computer explore a large range of potential options, states, and mathematical fits to find the "best" relationship between input and output? This approach can use a variety of statistical methods to describe trends or patterns in data or can use machine-learning methods to find correlations or behaviors in past data to predict or categorize future observations.

Some tasks are "easy" for humans, but difficult to program a computer to do (e.g., recognize a human face or emotion through body language). Artificial intelligence can use vast amounts of data to identify characteristics of faces or postures to classify or identify individuals or emotions. These tools are now being used in space weather research to identify events (e.g., coronal mass ejection intervals in solar wind and IMF data), distinguish between instrumental noise and real signals, or tease out relationships between multiple parameters.

7.7.4 Logical Fallacies or Sagan's BS Detector

How to make and understand a logical argument or conceptual model is at the heart of science. Science attempts to understand the "why and how" of the physical world – why does the Sun have a magnetic field? How does the Sun's magnetic field influence the flux of cosmic rays at Earth? etc. To do this, science can be thought of as using four approaches: theory, observations, experiments, and computer modeling. The fourth approach is clearly a modern addition to science, but has become a distinct approach. These methods all contribute to the same problem and can motivate and inform each other. At each step of the way, science often makes tentative and provisional conceptual models to inform the next step of the investigation. For example, two of the questions raised by the first modern scientists to observe and write about the aurorae in the sixteenth and seventeenth centuries were "what are the aurorae?" and "how are they created?" In Chapter 1 some of these observational studies were described (when and where the aurorae were observed, and that the same aurora was observed in Italy and France); the model or theory was proposed that aurorae must occur high in altitude for distant observers to see them. This led to the next question: what was the altitude of the aurorae? This led to the triangulation experiments of Cavendish in the eighteenth century determining that they occurred about 100 km in altitude). In the twenty-first century, we have developed complex atomic and molecular photochemical models of the upper atmosphere that can simulate the aurorae, comparing them with ground and space-based observations to test not only physical characteristics of the aurorae (at what height are they), but also their formation mechanisms. Note that each method or approach has limitations, assumptions, and uncertainties that make each new interpretation provisional or tentative since we recognize that our assumptions may not be valid, that further observations can invalidate our current theory, or that our model may not include all the relevant parameters, physics, or parts.

One significant issue for many explanations, models, or theories is that there can be multiple explanations consistent with the observations, theory, and current understanding. This requires scientists to develop testable hypotheses and experiments to attempt to distinguish the strengths and weaknesses of the competing ideas and models.

Unfortunately, two issues arise in the competition between models or ideas, in a general context and in science. The first is that science is a human endeavor practiced by people. We tend to equate our ideas or models with ourselves, and others' ideas or models with their creator (we personalize the model). So instead of simply arguing (hopefully in the scientific sense) the merits and weaknesses of the model, we argue with the modeler (instead of saying "the model is wrong," we say "you are wrong"). The modeler does not interpret that the model is being examined and criticized, but feels that they are being criticized. This leads to scientific debates or fights that become personal and often linger for much longer than they should.

The second issue is that, as humans, we are susceptible to biases that are part of our conceptual understanding or framework of reality. Many of these biases (like confirmation bias – remembering or including only examples or observations that fit your preconceived idea) or logical fallacies (like misunderstanding or misinterpreting small-number statistics) contribute to scientific arguments. Of course, scientists of good will and clear understanding of their limitations can still develop competing interpretations of the same phenomenon, event, or data set – and that is the heart of science.

So how do you create and understand scientific models, theories, and understanding? One tool you should have is the ability to create strong arguments and recognize weak arguments. Many weak arguments are built upon what are called "logical fallacies." These logical fallacies are easy to recognize once you know what to look for, but unfortunately there is a long list of them. Some of the most common ones that appear in science are described here, but I highly recommend Carl Sagan's essay on "The Baloney Detection Toolkit" from which this section drew inspiration. Like Sagan's essay, we explore each fallacy in turn by defining the method and provide an example or two of its use in the context of science and college, rather than politics and religion, which Sagan primarily drew from.

Ad hominem (Latin "to the man") is the point made above that we often equate an idea or theory with the person who is making it and therefore attack the person instead of the idea. (Example: "Moldwin is just an inexperienced data analyst and therefore we don't need to take seriously his comment on instrumentation" – a paraphrased version of an attack on me when I was a post-doc from a senior colleague in the field at a meeting, after I pointed out issues with one of the senior colleague's recent papers.)

Argument from authority is using power differential to justify an argument instead of the merits of the idea. (Example: "As the principal investigator of this mission, I know what I am doing, so trust me on my experimental plan." Instead of allowing others on the team to evaluate the strengths and weaknesses of the plan, the principal investigator relies on their authority to limit examination.)

Argument from adverse consequences is the justification for a plan or action by warning that (often unrelated) bad things will happen otherwise. (Example: "If we do not fail this engineering student now for turning in their assignment late, it will encourage other students to cheat on their exams.")

Appeal to ignorance is the fallacy that whatever has not been proved false must be true or whatever has not been proved true must be false. (Example: "Since there is no evidence of other life in the universe, we must be the only life in the universe.") The central premise is that uncertainty or ambiguity of the evidence often does not allow firm conclusions (absence of evidence is not evidence of absence). Those who promote the status quo often promote uncertainty (see Oreskes and Conway's *Merchants of Doubt* for how the tobacco and oil industries used this fallacy to promote the use of cigarettes and fossil fuels).

Begging the question, or **assuming the answer** is the fallacy that states as fact an interpretation of cause and effect. (Example 1: "We must not allow NASA space mission cooperation with China otherwise China will overtake US leadership in space." But does international collaboration diminish the US space program? Example 2: "This is an excellent scientific paper because it has many well-known authors." Can well-known authors write terrible papers?)

Observational selection, or cherry picking the "good" and discarding the "bad." (Example: A study highlights the data that support the hypothesis, but neglects or discounts the data that are contrary or inconsistent with the hypothesis.) Many "statistical" studies show a few "representative" examples that highlight the "best-typical" events and do not show events that are inconsistent with the generalization or hypothesis drawn from the study.

Statistics of small numbers – a blend of misunderstanding the nature of statistics and observational selection. (Example 1: "All five substorm events that occurred during our observational campaign showed a relationship with a polarity change of the IMF, therefore all substorms are driven by IMF polarity changes." Example 2: "All three substorms that occurred today have IMF polarity changes. Therefore we will be able to predict the next substorm by just looking for a polarity change.")

Misunderstanding of the nature of statistics. (Example: Marveling that the onset of large geomagnetic storms occurs on weekends nearly 30% of time [Saturday and Sunday are 2 out of 7 days of the week, i.e., 28%], or that half of all substorms occur within a week of Full Moon.)

Inconsistency (Example 1: Attributing the incredible technological successes of NASA to the excellence of the US educational system, while neglecting that US students perform near the bottom in many science and math assessments compared to the rest of the world. Example 2: Considering it reasonable for the universe to continue to exist forever into the future, but judging absurd the possibility that it has infinite duration into the past. My PhD advisor's (Jeff Hughes) advisor's (James Dungey) advisor (Fred Hoyle) coined the term "Big Bang" as a derisive name because he did not like to think of a beginning or "Genesis" moment. He developed a "steady state" cosmology that has not fared well observationally compared to the Big Bang theory.)

Non sequitur – Latin for "It doesn't follow." (Example: "Alien visitation must have happened, otherwise how do you explain the Egyptian pyramids?") Often those falling into the non sequitur fallacy have simply failed to recognize alternative possibilities.

Post hoc, ergo propter hoc – Latin for "It happened after, so it was caused by." (Example 1: "We were never concerned about global warming until scientists started studying climate using satellites, therefore satellites must be causing the warming." Example 2: "The resignation of Richard Nixon was precipitated by NASA canceling the Apollo Program.")

Excluded middle, or **false dichotomy** – considering only the two extremes in a continuum of intermediate possibilities. (Example: "Either model A is right or model B; they cannot both be correct." Of course both could be able to explain the observations equally well, or the phenomenon being explained could have two different causes.)

Short-term vs. long-term – a subset of the excluded middle or false dichotomy. (Example: "Musk, Branson, and Bezos should spend their fortunes on solving problems here on Earth instead of developing spaceships." Adapted from Sagan's example and recognizing that spaceships can be used to solve problems on Earth by carrying Earth-observing satellites to understand climate, severe weather, and agriculture, as well as enabling new communication technologies that can benefit global populations. In addition, billionaires, companies, or governments can allocate resources that have well-defined immediate impacts or invest in technology development for long-term objectives at the same time.

Confusion of correlation and causation. (Example: "Many climate variables have decadal periodicities, clear evidence that the 11-year sunspot cycle controls the weather.") See Section 9.8.2 for further explanation.

Straw man – creating an over-simplified or incorrect analogy to make it easier to find fault in someone's argument. (Example: "Why spend money on trying to predict space weather when we can't even predict if it will rain tomorrow.")

7.8 Problems

7.1 The Earth's magnetic field can be approximated as a dipole magnetic field (see Section 4.4) that is tilted about 11 degrees from the spin axis of Earth.

    (a) Create a model of the geosynchronous orbit magnetic field that a satellite would see at any magnetic longitude over a 24-hour period.

    (b) Plot the value of the magnetic field at three different longitudes (0, 45, and 90 degrees) over a day.

    (c) Describe why the sign of the radial component of the magnetic field changes over the day.

    [The dipole magnetic field can be broken down into components: the radial (B_r) and the north–south (B_θ) directions. The radial component points from the center of the Earth, while the north–south component is positive northward and is orthogonal to the radial component.

    The equations of a dipole field in polar coordinates (r, θ) are

- $B_r = 2M \cos \theta / r^3$,
- $B_\theta = M \sin \theta / r^3$,

where M is the dipole moment, which can be positive or negative. For Earth, $M = -8 \times 10^{15}$ T m^3 or $-31\,000$ nT r_E^3. The angle θ is the colatitude measured from the pole to the equator. It is often useful to use latitude (λ) instead of colatitude (θ); then the equations are

- $B_r = 2M \sin \lambda / r^3$,
- $B_\lambda = -M \cos \lambda / r^3$.]

7.2 A HEO satellite with apogee of 15 r_E slowly precesses in local time (LT) over its lifetime (the LT of apogee rotates). It can measure crossings of the magnetopause (see Chapter 4) and obtained the following data set:

[LT, r_E]
06, 16
07, 15
08, 14
09, 13
10, 12
11, 11
12, 10
13, 11
14, 12
15, 13
16, 14
17, 15
18, 16

(a) Create an empirical model that describes the location of the magneto-pause as a function of local time. (You can use Excel, Matlab, Python, or determine "by hand" a function that describes the data.)

(b) Compare your best-fit analytic function to that of an ellipse, with the foci at the center of the Earth and the distance to the subsolar point (magnetopause at local noon) the semi-major axis.

7.3 Use a similar approach to Problem 7.2, but use part of the NASA Spacecraft magnetopause crossing database,
https://omniweb.gsfc.nasa.gov/ftpbrowser/magnetopause/Intro.html
focusing on dayside crossings of the magnetosphere. First download the appropriate data set and bin the data by local time.

(a) What does the distribution of locations of the magnetopause at noon look like for your data set? (Create a histogram of all the crossing locations that occurred within an hour of noon.)

(b) Plot the noon local time bin magnetopause location as a function of upstream solar wind dynamic pressure. What type of relationship is found?

(c) Repeat for other dayside local time bins.

7.4 Research some of the limits of modeling in order to predict the behavior of a complex, coupled system such as the magnetosphere and write a short essay describing why we will never be able to forecast the physical state of the system perfectly.

7.5 Find a recent news story of a political argument regarding a scientific topic that uses a logical fallacy. Provide the URL to the story and describe what fallacy is being used.

7.6 Create a concept map of the different types of models, being sure to connect to specific space weather models.

Chapter 8
The Perils of Living in Space

"My God, Space is radioactive!" Ernie Ray's exclamation after seeing the data returned from Explorer 1, the first US satellite launched in 1958. The radiation was space radiation in the Van Allen Radiation belts.
   (Quoted in Hess, 1968)

8.1 Key Concepts

• ionizing radiation
• protection by Earth's atmosphere and magnetosphere
• rems (radiation equivalents in man)
• types of radiation: corpuscular and electromagnetic

8.2 Learning Objectives

After actively reading this chapter, readers will be able to:

• list the different types of ionizing radiation and describe how radiation impacts human biology;
• summarize the role of the Earth's atmosphere and magnetosphere in protecting both humans and technological systems from ionizing radiation;
• organize the different biological impacts on humans during space travel and construct possible plans for human exploration of the Moon and Mars that minimizes the impacts of space hazards.

8.3 Introduction

Life on Earth has an over 3.5 billion year history. It had a beginning and it will have an end. The longest that life can possibly exist on Earth into the future is approximately 4 billion years, though a number of catastrophes can happen on much shorter timescales (some of these are discussed in Chapter 9). In approximately 4 billion years, our Sun will run out of nuclear fuel and enter what is called the Red Giant phase of stellar evolution. It will expand perhaps past the orbit of Earth, vaporizing Mercury, Venus, and Earth. If humans are to survive the end of Earth, we will have to develop the technological capability to move to

143

another star system. Humans have made the first tentative steps off Earth. If you were born after November 2001, there have been humans in orbit on the International Space Station (ISS) your entire life. The USA, China, and even the commercial company SpaceX plan to have a manned presence on the Moon and Mars in the relatively near future. However, the technological and biological obstacles to space travel are daunting – some say even insurmountable.

On the surface of Earth, we are protected by the atmosphere and magnetosphere from the deadly electromagnetic and corpuscular radiation coming from the Sun and outer space. If we leave the **protection of Earth's atmosphere and magnetosphere** we will have to take protection with us. In addition, our bodies are not adapted to the extreme temperatures, the extreme radiation, the intense vacuum, the lack of Earth's gravity, and impacts from high-velocity micro-meteoroids that exist in space. This chapter describes how the space environment impacts living things (especially humans) and our current efforts to live and work on the surface of the Moon and Mars and ultimately around another star.

8.4 Radiation

There are two main **types of radiation: electromagnetic and corpuscular**. Electromagnetic (EM) radiation consists of massless particles of pure energy called photons. These photons also act as waves and so they have corresponding wavelengths and frequencies. Wavelengths describe the distance between the crests or troughs of a wave and frequency tells how many times a wave passes a point each second. The entire electromagnetic spectrum is composed of radio, microwave, infrared, visible, ultraviolet, X-ray, and gamma ray radiation, going from long wavelength to short wavelength. Figure 2.7 shows the entire EM spectrum schematically. Photons exist at different energies with radio waves having the lowest energy and gamma rays having the highest energy. Energy and frequency are related in a simple relationship,

$$E = hf,$$

where E is energy, h is the Planck constant, and f is the frequency. Frequency and wavelength are related to the propagation speed of electromagnetic radiation – that is, the speed of light in a vacuum. The relationship is $\lambda f = c$, where λ (the Greek letter "lambda") is wavelength and c is the speed of light.

Radio waves have the longest wavelengths, the lowest frequencies, and lowest energies, while gamma rays have the shortest wavelengths, the highest frequencies, and the highest energies. Visible light is made up of multiple frequencies that we perceive as different colors. Red has the longest wavelength and hence the lowest energy and frequency, while violet has the shortest wavelength and therefore the highest frequency and highest energy.

EM radiation from the Sun is mostly visible light, but light at essentially all wavelengths is also emitted. Much of the radiation does not reach the ground because of atmospheric absorption and reflection, particularly the high-energy photons such as UV and X-rays. However, in space the full intensity of sunlight is felt without the protection of the atmosphere.

Corpuscular or particle radiation is primarily subatomic (protons and electrons) or atomic or molecular particles. A natural background radiation environment comes from Earth, not only from space. Many elements are radioactive, meaning that they spontaneously decay from one element to another element by releasing different types of radiation. Madame Marie Curie[1] won Nobel Prizes in Physics and Chemistry for her work in understanding radioactive decay processes. In radioactive decay processes, both electromagnetic and corpuscular radiation can be released. Alpha (helium nuclei) and beta (electrons) are two common types of radioactive decay by-products. These are also part of the solar wind and cosmic ray populations, though their origin is the ionization of helium and other gases, not from radioactive decay.

8.5 Biological Impacts of Ionizing Radiation

Both EM and corpuscular radiation can be **ionizing radiation** – that is, radiation that carries enough energy to ionize an atom or molecule. In biology, this radiation can also interact with living cells by damaging or destroying the cell or DNA contained in the cell. The building blocks of all life on Earth are cells. Cells consist of mostly water and the elements hydrogen, carbon, nitrogen, and oxygen with a little phosphorus and sulfur. Radiation, both corpuscular and high-energy EM radiation, can directly interact with these elements by removing or exciting an electron and hence raising the internal energy level or ionizing the elements. This dramatically changes the chemical reactivity of the atom. Reactivity is the probability or likelihood that the atom or molecule will combine with another or take part in a chemical reaction with

---

[1] Marie Curie (1867–1934) and Pierre Curie (1859–1906). Madame Curie was a Polish-born French physicist who along with her husband shared the 1903 Nobel Prize in Physics with Becquerel for their work in understanding radioactivity. Madame Curie went on to study the chemical and medical applications of radioactivity and won the 1911 Nobel Prize in Chemistry for her work. A unit of radioactivity (the curie, Ci), named in their honor, is equal to 3.7×10^{10} decays per second.

another atom or molecule. Molecules containing excited or ionized atoms can react with other cells in a way that is detrimental to the living organism. Depending on the type and intensity of the radiation, the organism can suffer a wide range of health effects. For humans these can include reduction in white blood count, nausea and hair loss, development of cancer, or immediate death. Cells that are most sensitive to radiation include white blood cells and the cells that make the white and red blood cells. Therefore radiation exposure can have significant impacts on the body's immune system. The intensity of the radiation is very important – not only the absolute amount but also the energy. Very energetic particles can impact a large number of molecules, causing significant ionization. High-energy EM radiation reacts differently with living tissue than energetic particles. EM photons lose all of their energy with a single interaction. In contrast, energetic particles lose their energy through collisions with large numbers of molecules. However, EM radiation (particularly X-rays and gamma rays) can create secondary electrons and photons, which can then interact with nearby atoms creating more secondary electrons and photons, etc. To distinguish the two different processes, particle radiation interaction with living tissue is called direct ionization radiation, while the process involving electromagnetic radiation is called indirect ionization radiation. This is because the same particle can directly interact with a large number of atoms or molecules, whereas a photon directly interacts with only one atom or molecule, but indirectly affects others by producing other photons or electrons.

One possible outcome of the interaction between both types of radiation is that, instead of ionizing the atom, the radiation knocks an outer electron into a higher energy level, creating what is called an excited-state atom or free radical. Free radicals are highly reactive and can damage surrounding molecules. Since living tissue contains mostly water molecules, one of the most common free radicals produced is excited hydroxide (OH^*, where the star or * is the symbol used to indicate that it is in an excited state, the same symbol used to distinguish an excited atom or molecule discussed in Chapter 5 with regard to aurorae and photochemistry). Hydroxide, a strong oxidizing agent, can cause abnormal chemical reactions in living cells.

Free radicals can rupture cell membranes causing the destruction of the cell. If enough cells are killed, the biological function associated with the cells (i.e., white and red blood cells, internal organs) could cease. Death can then occur either due to the loss of the organs' functionality or from infections that overwhelm the system due to the shutdown of the body's immune system (decreased white blood cell counts). Another way that radiation can damage living cells is direct interaction with complex molecules such as proteins and nucleic acids (which make up the genetic DNA). The radiation can break DNA strands or break apart the protein preventing their proper functioning. DNA has the ability to repair itself, and because much of the genetic code is highly redundant, damage at a few sites is not detrimental. However, if the intensity of radiation is great enough, the living cells' ability to repair themselves could become overwhelmed and cause permanent damage. This damage could lead to cancer, genetic mutations in offspring, or decreased cell function.

The outcome to a biological system exposed to radiation depends on the type of radiation (particle or EM), the energy of the radiation, the amount of radiation, and the time period of exposure. For humans, various types of radiation impact different organs differently and there are differences between the sensitivities to radiation among men and women. Most electromagnetic radiation is benign to humans, though high-energy EM radiation (UV, X-rays, and gamma rays) can be dangerous or lethal. Most of you have had experiences with medical X-rays, often at the dentist's office. The dental technician covers you with a heavy lead apron and then leaves the room when the X-ray machine is turned on. These precautions are made due to the potential long-term damage that X-ray exposure can have on living tissue and the high penetration power of X-rays. We also are taught to wear sunscreen and hats when we are out in the sun to prevent sunburn. Sunburn is direct damage to the skin by solar ultraviolet radiation. Too much exposure can cause short-term pain and swelling and also lead to long-term health effects such as skin cancer. About three million Americans are diagnosed with non-melanoma skin cancers each year and about 2000 Americans are killed by this disease each year. Melanoma, the most deadly form of skin cancer, kills an estimated 8000 Americans a year. The direct cause of this cancer is UV damage to the largest organ of the human body – the skin.

A source of possible confusion regarding radiation is measuring how much radiation there is and how to measure the effect of the radiation on material or living organisms. One way to describe the level of radiation is using the units curies (Ci) and becquerels (Bq). Henri Becquerel[2] discovered spontaneous radioactive decay and the Curies (wife and husband team) explained and measured it. All three shared the 1903 Nobel Prize in Physics for this work. The Ci and Bq

---

[2] Antoine Henri Becquerel (18520–1908), French Nobel-Prize-winning physicist who discovered radioactivity. The SI unit of radioactivity (becquerel, Bq), named in his honor, describes the number of radioactive decays of a certain amount of material per second (therefore the larger the number of Bq, the more radioactive the substance).

measure the number of radioactive decays per second in different radioactive materials (such as uranium). This describes a property of the activity of the source of the radiation, but does not say anything about the type or effect the radiation has on any material or biological system.

Rads and grays[3] (Gy) are used to measure the amount of energy absorbed by a specific material. They are given in units of energy per mass, or joules per kilogram in SI units. Sieverts (Sv) or **rems (radiation equivalents in man)** measure what are called dose equivalents, taking into account the effects different types and energies of radiation have on human tissue. The units of sieverts are also joules per kilogram. Rems or sieverts are the units used when estimating radiation dose and effects on humans.

What happens to a human exposed to higher and higher doses of radiation? A complicating factor is that equal exposure to different types of radiation does not lead to equal biological impacts. A gray of alpha particles (helium nuclei that are by-products of radioactive decay and the second most abundant ion in the Sun and solar wind) does not produce the same effect as a gray of beta (electron) radiation or gamma-ray radiation. That is because human tissue absorbs various types of radiation differently. Therefore, the units of sieverts or rems are used when discussing human biological effects of radiation because these units take into account the quality factor of the different types of radiation with respect to their impact on living tissue.

Typically an American is exposed to 0.365 rem per year. This dose comes from a variety of sources, as shown in Figure 8.1. The biggest source of radiation is from naturally occurring radon gas from the radioactive decay of radium, which has as its source uranium-238. These sources occur naturally in Earth's crust. Radon itself has a short half-life and decays into radioactive polonium, which is an alpha-emitter. Radiation from naturally occurring radioactive isotopes inside your body provides the next highest dose.

Exposure to high doses of radiation can lead to acute radiation effects (in medical terminology, acute means sudden or lasting a short time, in contrast to chronic, which is long-developing). At an exposure level of 25 rems, subtle, hard-to-detect reduction in white blood cell counts (WBC) can occur. At 50 rems, reduction in WBC is easily detectable. WBC returns to normal after a few weeks. White blood cells are an important component of the body's immune system.

---

[3] Louis Harold Gray (1905–1965), British physicist who was instrumental in the development of radiation biophysics (radiology). The SI unit of radiation dose (gray, Gy) is named in his honor.

Average Annual Radiation Exposure of Americans
365 mrem

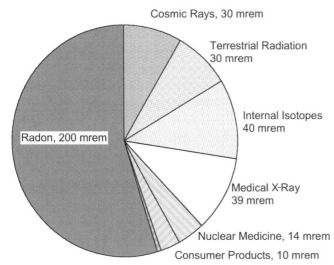

Cosmic Rays, 30 mrem

Terrestrial Radiation
30 mrem

Internal Isotopes
40 mrem

Radon, 200 mrem

Medical X-Ray
39 mrem

Nuclear Medicine, 14 mrem

Consumer Products, 10 mrem

Figure 8.1 Sources of radiation exposure for a typical American. The unit of radiation exposure is millirem (0.001 rem = 1 mrem). (Adapted from data from the US Department of Energy)

Therefore reduction in WBC can lead to death if there are other illnesses or infections. At 75 rems there is a one-in-ten chance of nausea. Nausea is a symptom of radiation sickness because crypt cells that line the intestine are especially sensitive to radiation. Damage to these cells can trigger nausea, vomiting, and dehydration. At 100 rems there is also a 10% chance of temporary hair loss. These two symptoms (nausea and hair loss) are often associated with cancer radiation treatment, which takes advantage of the ability of radiation to penetrate healthy bone and tissue and be absorbed by and destroy cancerous cells. Even though the radiation is highly "focused," it not only impacts cancerous tissue, but also can damage surrounding healthy tissue. At 200 rems there is a 90% chance of radiation sickness and moderate WBC reduction. At 400 rems there is a 50% chance of death within 30 days; 600 rems is lethal to most people within 3–30 days. Exposure to over 10 000 rems is almost instantaneously lethal, with death coming in less than a day. Unfortunately, human understanding of health radiation effects at these high limits has come from studying the aftermath of the Hiroshima and Nagasaki atom bomb attacks and the 1986 Chernobyl Nuclear Power Plant disaster.

What types of exposure are humans subjected to in space? An astronaut in low Earth orbit such as on the space station receives about 0.1 rem per day (or a typical American's yearly dose in under 4 days). For comparison, a chest X-ray gives an exposure of 0.010 rem (10 millirem); therefore, an astronaut's exposure is equivalent to about

10 chest X-rays per day. If an astronaut is on a spacewalk during a solar storm, the dose at LEO could be 10 to 1000 times higher than the normal background. Therefore astronauts need to be concerned about solar storms, even within the protective cocoon of Earth's magnetosphere. For trips to the Moon and Mars, large solar energetic particle events from the Sun can give lethal doses to unprotected astronauts, and the chronic exposure to galactic cosmic rays outside the Earth's protective magnetosphere can lead to a variety of negative health effects.

During the Apollo missions to the Moon, the US space program and the astronauts got lucky. In the 1970s, radiation effects from the Sun were not fully appreciated. Apollo 16 was launched in April of 1972. Apollo 17 (the last of the six lunar landings) was launched in December. In August of 1972 one of the largest solar proton events ever measured occurred. If the astronauts had been walking on the Moon during this event, there is estimated to be a 50–50 chance that one of them (Eugene Cernan or Harrison Schmitt) would have received a lethal dose.

Besides acute radiation effects, there are effects of long-term or chronic radiation exposure. Effects of chronic radiation exposure are usually slow to develop, with latencies (medical term for time period) of effects to develop usually measured in years if not decades. These include genetic defects in offspring and the development of cancer. These effects are characterized by the total ionizing dose (TID, see Section 6.5.3) exposure, which is just the sum of all the radiation received over a certain timeframe. There are annual and career limits on the total radiation exposure that a radiation worker is allowed. The Occupational Safety and Health Administration (OSHA) in the United States provides these limits for those who work in high-radiation environments, such as radiation technicians, airline pilots, nuclear power plant operators, and astronauts. The number of radiation workers has increased in recent years because of such things as increased luggage X-ray inspection at airports and the use of ionizing radiation on the US mail to neutralize biological agents such as anthrax.

The career limits depend on gender and age, with men and older workers allowed exposures to higher doses. Developing embryos and fetuses are especially vulnerable to radiation effects, and therefore more stringent exposure limits are placed on women. For a 25-year-old woman, the career dose exposure limit is 100 rems. It is 150 rems for men the same age. These limits essentially increase by a factor of 3 to age 55.

An astronaut who spends three months on the International Space Station receives a typical dose of less than 10 rems. Spending the same amount of time working inside a laboratory on the surface of the Moon would give similar exposure rates. However, these exposure rates do not include effects of solar storms or possible exposures during spacewalks. Astronauts are particularly vulnerable when they are outside the protective shielding of their spacecraft or lunar base. A three-year mission to Mars (two-year round-trip flight time and one year on the surface) would give approximately 100 rems due to typical galactic cosmic ray background radiation. Again, with a few solar storms, astronauts could be exposed to lethal doses of radiation.

In addition to these health effects, radiation exposure can have cognitive and neurological impacts. In tests on mice exposed to typical radiation doses expected for a six-month Mars trip, the mice developed decreased memory and learning abilities and increased anxiety. Currently, providing radiation protection to astronauts is one of the biggest technological hurdles to be overcome before manned missions to the Moon and Mars could become possible. So in one sense the biggest hurdle to interplanetary space travel is the limits of our biology.

How can we shield astronauts from radiation? The ability to penetrate matter depends on the type and energy of the radiation. Low-energy alpha particles can be stopped relatively easily by a sheet of paper or a shirt. Energetic particles – especially energetic electrons or killer electrons – can penetrate deeply through most matter. Spaceships are usually designed with thin metal (often aluminum) walls to save weight. Astronauts on the ISS are instructed to go to the center of the space station during space storms to provide the maximum protection available from the spacecraft structure.

There is a proposal to fly a large magnet with a spaceship to Mars. The magnetic field would then act as a shield against some of the lower-energy energetic charged particles, though there are significant technological hurdles in creating a large enough magnet or electromagnet to make this idea feasible. Although Mars does not have a global intrinsic magnetic field, there are pockets of remnant magnetic field that act as mini-magnetospheres. Therefore potential human landing spots could use the remnant magnetic field near the surface to protect against some of the background radiation. Another proposal is to build subsurface labs in order to allow the lunar or Martian "dirt" (called the regolith) to act as shielding. The discovery of lava tubes on the Moon and Mars provides an opportunity to potentially make subsurface habitats relatively quickly, though there has not been any exploration of the subsurface caves. Lava tubes also exist on Earth and are formed during fast basaltic lava flows where the exterior cools before the subsurface flow and forms a crust.

8.6 Problems of Long-Duration Space Travel

Besides radiation, there are a number of potentially fatal space environment impacts on human exploration of space. These include the vacuum of space, reduced gravity, and micro-meteoroids. In addition to these health and safety issues, new research has found that radiation can cause neurological changes that can impact performance of long-duration space explorers. Each of these issues will need to be addressed for humans to work and live in the space environment.

8.6.1 Vacuum of Space

Humans living at sea level are subject to an atmospheric pressure of 1 atmosphere (atm) or 1 bar (b), or in SI units about 100 000 pascal (Pa). This pressure, due to the force of the atmosphere pushing down on the surface, is equivalent to 14.7 lb per square inch (or about 1 kg per square centimeter). If an astronaut is exposed to the vacuum of space, death will usually result unless she can be rescued within approximately 90 seconds. Within approximately 10 seconds the astronaut will lose consciousness. If the exposure to vacuum is sudden, problems with explosive decompression (the rupture of lungs if astronaut attempts to hold their breath) can occur. There have been several instances where astronauts, pilots, and high-altitude parachutists have experienced rapid decompression. One issue of rapid decompression is decompression sickness, more commonly called "the bends." This is the same problem that deep-sea divers need to be concerned about. Decompression sickness is caused by gases being released from the blood stream and tissue. (The effect is similar to opening a bottle of seltzer water. The carbon dioxide gas is dissolved in the water under high pressure. When you open the bottle quickly the gases are released from the solution and cause the water to fizz out.)

Some myths perpetuated by Hollywood (such as in Arnold Schwarzenegger's movie *Total Recall*) are that your body will explode and your blood will boil if you are suddenly exposed to the vacuum of space. Though bad things occur, they aren't that dramatic. The release of water vapor in your tissue can cause you to swell to up to twice your normal volume (unless constrained by a pressure suit). Arterial blood pressure will fall within a minute and effectively shut down blood circulation. After the initial rush of air from the lungs, continued release of air and water vapor will cool the nose and mouth to near freezing due to the expansion of the gas. Death will come within 90 seconds. Decompression can happen for astronauts if their spaceship, spacesuit, or lunar or Martian habitat gets breached, and therefore designs for these suits and structures must incorporate features that will prevent or fix any leaks.

8.6.2 Reduced Gravity

Another physiological issue that astronauts face when traveling in space is the greatly reduced gravity environment. On the surface of Earth we are subject to an acceleration of 9.8 m s^{-2}. This gravitational acceleration of Earth is written as g. The force with which Earth pulls down on an object on its surface is equal to this acceleration times the mass of the object (force = mass × acceleration). We call this force "weight" when referring to measuring how much force an object exerts on Earth due to its mass. It is also often referred to as a g force. So if a roller coaster subjects you to 4 gs, you are subject to a force four times the normal force of gravity. As you move away from Earth, the acceleration due to the planet decreases since the gravitational force is inversely proportional to distance between two objects raised to the second power (or squared). Also if you are in orbit around Earth, you are constantly accelerating due to your motion (called centripetal acceleration). You are "free-falling" toward Earth along with your spacecraft. Therefore, there is no gravity relative to your surroundings. In a sense, you become "weightless." If you were at the altitude of the International Space Station (about 300 or 400 km) and not in orbit around Earth, the force of gravity (which is only down about 10% from its surface value) would pull you back to Earth quickly. Astronauts in low Earth orbit experience micro-gravity because they are orbiting at such a speed that they continually "free-fall" around Earth.

The human body is designed to live in a one-g gravitational environment. If you live on the space station or a spaceship going to Mars, your muscles and bones are not subject to the normal stresses of living on Earth and can begin to atrophy. An astronaut can lose up to 1–2% bone density each month, and even more muscle mass. After a half-year mission, an astronaut can lose over 40% of muscle mass and over 10% of bone density. This makes astronauts very weak in Earth's one-g environment and could even lead to difficulty on Mars (1/3 g gravity). If astronauts spend significant amounts of time on the Moon (1/6 g gravity) or Mars, muscle and bone loss could have serious impacts on the success of the mission. Imagine being an astronaut landing on Mars after an eight-month trip and not being able to stand up against the Martian gravity because of muscle loss suffered on the journey. Astronauts currently spend much of their time on the ISS exercising (they are allocated about 2 hours per day) in order to attempt to slow the effect of low gravity on muscles and bones. Several spaceship designs, such as in Arthur C. Clarke's *2001: A Space Odyssey* or *Rendezvous with Rama*, incorporate artificial gravity. The outer wall of a spinning spaceship would exert centripetal force on an astronaut. If the ship is large enough and spins fast enough, the one-*g* environment of Earth could be mimicked.

In addition to impacts on bone density and muscle mass, new research has found micro-gravity impacts on cardiovascular health. A recent study found that many astronauts on the ISS have stagnant or reverse blood flow in a major upper-body blood vessel that can cause a clot. This can lead to the clot traveling to the lungs, causing a pulmonary embolism, which is an extremely dangerous condition. As humans spend more and more time in orbit, we learn of new and often surprising health impacts.

8.6.3 Micro-meteoroids

Micro-meteoroids are present in the solar system and can be from comets or asteroids. On a typical day about 33 metric tons (73 000 lb) of meteoroids strike the Earth. Though these can be very small (the size of a grain of sand or smaller), they can have large relative velocities compared to an astronaut in LEO or on the surface of the Moon. These velocities can be tens of thousands of miles per hour (tens of m s^{-1}) so even a speck of dust can have a tremendous impact on a spaceship or astronaut. A grain of sand moving at 17 000 miles per hour has about the same kinetic energy as a bowling ball moving at 60 miles per hour. Impacts of meteoroids on the Moon have been observed by telescopes as bright flashes, and 200 new craters with sizes 3–43 m (10–140 feet) were observed during a period of about five years by NASA's Lunar Reconnaissance Orbiter (LRO). Impacts of meteoroids with a mass of a few kilograms create small craters a few meters across, while the larger craters imply the Moon is hit by faster and larger meteoroids as well. To avoid a Moon base being hit, there is little that can be done except perhaps developing advanced radar detection systems, building the laboratories deep inside existing craters, and hardening their construction to survive impacts. Building them underground would protect astronauts not only from meteoroid impacts, but also from some of the space radiation.

8.6.4 Cognitive and Neurological Issues

Recent studies on astronauts and mice have shown that background space radiation exposure has potential neurological and cognitive impacts. Identical twin astronauts (Mark and Scott Kelly) participated in an extensive inter-disciplinary study looking at the effects of space flight on humans including physiology, genetics, and neurology aspects.

Scott Kelly spent one year on the ISS while Mark was on Earth. Some of the findings from the study found changes in Scott's gene expression, with much of the change consistent with radiation damage of the DNA. During the one-year mission, cognitive tests showed little appreciable change, but immediately after landing, and persisting for six months, was a significant change in results of tests on cognitive speed and accuracy. It was hypothesized that this may be due to re-adjustment to Earth's gravity, and if so could create issues for Mars missions as well. More recent research on mice exposed to six months of radiation, which is expected on a trip to Mars, found cognitive and neurological changes. The mice exposed to radiation had impacts on learning and memory and increased anxiety compared to the non-exposed control group. This could have significant impact on the performance of a crew during critical and highly dangerous landing and initial surface deployment. It also could have long-term consequences for long-duration human exploration.

8.7 Living on the Moon and Mars

Astronauts would face a myriad of potential health issues traveling to and working on the Moon and Mars. These are on top of the estimated cost of $1 million per kilogram to put something on the surface of the Moon (so every liter bottle of water brought to the Moon costs $1 million to get it there). NASA has ambitious plans to return to the Moon to stay and use the Moon as a gateway to Mars. Elon Musk, the owner of the commercial rocket company SpaceX, has made sending large numbers of humans to live on Mars a life goal. Mars is currently the more problematic destination because a round trip would be of several years' duration. A possible solution is to plan a one-way trip! Although some explorers and planetary scientists have volunteered, the ethics, and potential political and legal issues involved, do not allow such an endeavor through a government-sponsored trip, but commercial companies such as SpaceX are proposing having astronauts live on Mars and not return to Earth. A mission of up to three years, or a permanent-presence mission, would require overcoming technical hurdles such as radiation and micro-meteorite protection, bringing enough fuel, food, air, and water (or the technology to extract these resources from Mars), overcoming the physiological effects of traveling and working in reduced-gravity environments, and finally the psychological stress involved in living in such a harsh environment without any chance of rescue. A tooth infection or broken limb could be catastrophic for a limited crew that depends on the entire team for survival. NASA, Europe, Japan, and India are currently sending space-craft and rovers to study the "red planet," search for water, and develop technologies to extract resources (fuel, oxygen, and water) from the Martian soil or from the polar ice caps. Rocket fuel is essentially hydrogen and oxygen and therefore the discovery of frozen water near the surface of Mars means that there is a ready source of ice to provide fuel, water, and breathable oxygen for the crew to live on Mars and return safely to Earth without having to bring everything with them from Earth. The physical health of the astronauts due to radiation and low gravity on both the Moon and Mars are the main issues for long-duration missions that currently do not have tenable solutions.

These radiation and potentially micro-gravity impacts on physical and mental health are on top of other psychological stressors that dangerous, isolated, and confined space travel induce. Being away from Earth, family, friends, and other humans while confined in a small space with a small number of other crew members can lead to behavioral issues that can impact the performance and mental health of the astronauts. NASA has developed technology and techniques and has a very rigorous astronaut selection process and training to help over-come some of these issues. However, all of the long-duration missions (up to one year) have been on the ISS, and much is still to be learned on how human behavior is impacted by multi-year trips in hostile, confined spaces far from home environments.

8.8 Interstellar Travel

Space weather occurs not only in the context of our solar system, but also around other stars and in the interstellar and intergalactic medium. So to travel between the stars would require the same technological protections needed to go to our closest cosmic neighbor, the Moon. The greatest challenge to interstellar travel for humans is the vast distances between the stars. To get to our nearest stellar neighbor, even traveling at the speed of light, would require a four-year journey. To reach the center of the Milky Way galaxy would take 30 000 years at the same speed. Our current technology is theoretically capable of going about 10% the speed of light and so it would still take at least 40 years to visit Proxima Centauri (and four years to send back a radio message to mission control that you made it). In 2016, astronomers discovered that there is an exoplanet orbiting Proxima Centauri ("exo" is Greek for "outside," and exoplanets are planets orbiting other stars). This exoplanet is named Proxima b and is within the "habitable zone," so liquid water could potentially exist on its surface.

Stars are constantly being born, emitting high levels of UV during the initial phases of their evolution; energetic particles are constantly being accelerated in huge blast waves during the death throes of stars; and cosmic collisions between stars and black holes fill the universe with their radiation. Other stars also show variability, in some cases much stronger and shorter timescale variability than our Sun. Therefore not only is any interstellar journey long, it is perilous. Hence, if there is life elsewhere in the universe, even if it has evolved into intelligent life, it is hard to convincingly argue that these beings will have the capability to visit our corner of the galaxy.

However, in the last century we have begun to beam EM radiation out into space from our radio and TV broadcasts, and in the last 35 years we have sent probes – the Pioneer and Voyager spacecraft – out into interstellar space. Voyager 1 and 2 left our heliosphere and passed into interstellar space in 2012 and 2018, respectively. Both will continue to fly across the galaxy for millennia. Who knows if they will ever come close to another star system that has a spacefaring civilization. They carry a plaque and recordings of the civilization that created the spaceship, with a star map pointing out our Sun and a map of our solar system pointing out Earth. Will someone find this, decipher it, and have the capability to come look for its owner? Will we still be here?

8.9 Supplements

8.9.1 Additional Learning Objectives

After actively reading these supplements, readers will be able to:

- describe the special relativity effects on time, length, and mass and calculate the change of these parameters for velocities that approach the speed of light;
- assess the current best estimates of each term in the Drake equation and create their own Fermi approximation.

8.9.2 Special Relativity

Special relativity describes the motion of objects that are moving close to the speed of light. The speed of light is 3×10^8 m s^{-1} (or 300 000 km per second). This is really fast compared to the normal speeds we are used to on Earth – a photon of light could circle Earth nearly 7.5 times in one second. Albert Einstein developed the special relativity theory during his *annus mirabilis* (Latin for "extraordinary year") of 1905. In that year, he published four papers (including his paper on special relativity), each of which is Nobel-Prize worthy, though only one (his theory of the photo-electric effect) was cited in his 1921 Nobel Prize announcement. The other two papers described the motion of particles suspended in a fluid (Brownian motion) and the equivalence of matter and energy, which included his famous $E = mc^2$ equation.

Special relativity describes electricity and magnetism for moving bodies. It explained the famous Michelson–Morley experiment, which found that the velocity of light is independent of Earth's relative motion and hence light does not require a medium to propagate (unlike sound waves). The original idea was that light propagated like a sound wave through a medium called the ether. The Michelson–Morley experiment attempted to measure the relative motion of Earth through this ether by looking at the propagation of light split into two beams moving orthogonal to one another. One beam pointed in the direction of motion of Earth in its orbit around the Sun and the other at right angles to this. Michelson and Morley found no difference in the velocity of the light, regardless of the direction of the light beam. The essential point of Einstein's paper was that there is not an absolute reference frame. In order for this postulate to work, Einstein needed to make the conjecture that the speed of light is independent of reference frame and moves at a constant velocity (called c) regardless of the motion of the observer or object emitting the electromagnetic radiation. Essentially he modified classical mechanics (the study of the motion of objects that was developed by Isaac Newton and therefore called Newtonian mechanics) so that it was consistent with Maxwell's electromagnetic theory.

This leads to some pretty interesting (or even bizarre) conclusions for objects moving very fast, including time dilation, length contraction, and the relativity of mass. Time dilation is the fact that a clock moving with respect to an observer runs slower than it does when it is at rest. Therefore a clock onboard a fast-moving spacecraft will run more slowly than the same clock on the ground, as observed by an observer on the ground. This has been demonstrated experimentally. Therefore time is relative. An astronaut traveling near the speed of light will "age" differently from those left on Earth. Imagine going on a round-trip interstellar journey at nearly the speed of light and coming back to Earth and finding that your children are now older than you! This completely non-common-sense conclusion was subject to intense scrutiny by scientists. Over the years, all of the main predictions of special relativity have been tested and confirmed. Special relativity is considered one of the most well-established concepts of physics.

Length contraction states that the length of an object in motion with respect to some observer appears to the observer to be shorter than its length L_0, when it is at rest with respect to the observer (L_0 is called the rest length). Length contraction occurs only in the direction of the relative motion. So this means that to an astronaut on a fast-moving spacecraft, objects on Earth appear shorter than they did from the ground (and to an observer on the ground, the spacecraft appears shorter than when it was waiting for lift-off on the ground).

The relativity of mass is another interesting consequence of special relativity. Objects moving at high speeds relative to an observer have larger masses than the same objects at rest. Therefore a relativistic electron in Earth's radiation belts is more massive than an electron at rest. In order to understand a relativistic electron's interactions with other particles and its dynamics, special relativity needs to be taken into account.

The magnitude of all three of these effects depends on the relative speed of the particle compared to that of light. This is expressed as a ratio v/c. When v/c is close to 1, special relativity effects are appreciable. Therefore objects with v/c close to 1 are called relativistic. The actual equations showing the magnitude of this effect are:

$$t = \left(t_0 / \sqrt{1 - v^2/c^2} \right) \text{(time dilation)},$$

where t_0 is time on a clock at rest relative to an observer, while t is the time on a clock in motion relative to the same observer. The symbol $\sqrt{}$ is the notation for square-root.

$$L = L_0 \sqrt{1 - v^2/c^2} \ \text{(length contraction)},$$

where L_0 is the length at rest relative to an observer, while L is the length of an object in motion relative to the same observer.

$$m = \left(m_0 / \sqrt{1 - v^2/c^2} \right) \text{(relativity of mass)},$$

where m_0 is the mass at rest relative to an observer, while m is the mass of an object in motion relative to the same observer.

Though it is not intuitive that the measurement of time, length, or mass should be dependent on the relative motion of the observer, these effects appear only when the relative velocity approaches the speed of light. For almost all motion that is observed on Earth and in the solar system, regular Newtonian mechanics works well. One of the big implications of special relativity is the cosmic speed limit of c. Note that as $v \rightarrow c$, the mass goes to infinity, length contracts to zero, and time stops. The implication of special relativity is that no object can go as fast or faster than the speed of light. This has tremendous implications for interstellar communication and travel, since distances between the stars are measured in light-years, and distances across the galaxy in 100 000 light-years.

8.9.3 Estimation and the Drake Equation

One of the tools of science is estimation. Critical thinkers are able to estimate essentially anything. Enrico Fermi[4] was famous for his ability to make quick and usually accurate estimates based on clearly stated assumptions but little or no data. These types of estimations are now often called Fermi approximations. One of the famous Fermi approximations is estimating how many piano tuners there were in Chicago. To start, estimate how many people lived in Chicago (P_c), how many people live in an individual household (H), what fraction of households had pianos (P_h), what fraction of them had their pianos tuned yearly (f), how long it would take a piano tuner to tune a piano (t), and how many hours a piano tuner worked each year (T). From this you can estimate how many pianos are tuned in a given year (population of Chicago / number of people per household × fraction of households with pianos × number of piano tunings per year) = number of pianos tuned in Chicago each year (or using our symbols $P_c/H \times P_h \times f$). You can also estimate the number of pianos an individual piano tuner can tune each year (number of hours per day worked × number of days per week worked × number of weeks per year worked × number of pianos tuned per hour). Then divide the number of pianos tuned each year by the number of pianos an individual piano tuner can tune each year, and you have an estimate of the number of piano tuners in Chicago.

The Drake[5] equation is a Fermi approximation that attempts to estimate the number of advanced technological communicable civilizations in the galaxy at any one time. It is a series of estimates of the number of star systems, the fraction of those that have habitable planets, the fraction of those that have life, the fraction of those that have advanced civilizations, and the average lifespan of a technological civilization. The Drake equation essentially addresses each of the

---

[4] Enrico Fermi (1901–1954), Italian-born American Nobel-Prize-winning physicist who developed the first controlled chain fission reaction, which led to the development of the atomic bomb and nuclear power. The element with atomic number 100, which was discovered the year after his death, was named in his honor (fermium).

[5] Frank Drake (1930–), American radio astronomer who in 1961 developed an estimate of the number of civilizations in the Milky Way galaxy and helped usher in the new field of astrobiology, the study of the origin and evolution of life in the universe.

questions involved with the Search for Extra-Terrestrial Intelligence (SETI). SETI asks the question, "Are we alone in the universe?" Since the development of radio astronomy, we have been able to scan the heavens looking for any form of radio communication that may have been beamed towards Earth. Because of the vast distances of even our nearest stellar neighbors, the radio signals left their source hundreds to thousands of years ago. In the Drake equation,

$$N = R^* \times f_p \times n_e \times f_l \times f_i \times f_c \times L,$$

N is the number of technical civilizations in our galaxy that can communicate with us,

R^* is the rate of star formation in our galaxy,

f_p is the fraction of stars that have planets,

n_e is average number of planets that can potentially support life per star that has planets,

f_l is the fraction of planets that develop life,

f_i is the fraction of planets with life that develop intelligent life,

f_c is the fraction of planets with intelligent life that are willing and able to communicate,

L is the expected lifetime of such civilizations.

Determining many of these parameters are active areas of research, especially the star formation rate in our galaxy and the search for exoplanets to estimate the fraction of stars that have planets. A new field of space science called astrobiology is attempting to address the origins and evolution of life on Earth and estimate how common it is elsewhere. The latter is being done currently by looking for evidence of life – past and present – on Mars and potentially some of the moons of the outer planets. Of course most of the parameters are not well constrained and therefore their values have large possible ranges (for all the "fractions" many are not constrained at all by current observations and can be anything from 0 to 1). Our single data point for life in the universe is ourselves. So we know that the probability is not exactly zero, but there are some scientists who believe that the conditions required over the last 4.5 billion years of our own development are so particular that intelligent life elsewhere is highly unlikely. Therefore estimates of intelligent civilizations in the galaxy range from 1 (us) to thousands. One difficulty with estimates of a large number of advanced civilizations existing at the same time in the galaxy is the fact that we haven't found each other yet. If advanced civilizations last for 100 000 or millions of years, why haven't we found evidence of them yet? This is often called the Fermi paradox because after hearing about early high estimates from the Drake equation, Fermi asked "Where are they?"

One of the largest unknowns as well as one of the most important parameters in the Drake equation is L, the lifetime of a civilization. The lifetime of our technological civilization capable of sending messages across space is just over 100 years old. In the last 60 years, we have begun to understand some potential man-made and natural disasters that would end our civilization (e.g., global nuclear war, impact by a large asteroid). So how long does a typical advanced civilization last? Are we alone?

8.10 Problems

8.1 What is the wavelength of an FM radio wave ($f = 100$ MHz)?

8.2 Name and describe two ways in which radiation is used for medical purposes.

8.3 Using a scale height (H) of 8 km, at what altitude does atmospheric pressure fall to 10 mbar?

8.4 At space shuttle altitudes, what is the strength of gravity (alt = 400 km, $g = 9.8$ m s^{-2})?

8.5 What is the mass of a radiation belt electron moving at (a) 0.9 c, (b) 0.99 c?

8.6 Estimate the number of college students in the United States. State your assumptions and list your approximations algebraically before solving numerically in a form similar to the Drake equation.

8.7 One way to reduce the exposure of radiation for trips to Mars would be to develop better propulsion systems to get to Mars quicker. This reduces the chronic (total dose) exposure as well as reducing the likelihood of exposure to an acute solar storm event. What are some of the other benefits of a "quick" trip to Mars for astronaut health compared to a nominal three-year mission?

8.8 Create of concept map of the biological hazards of space travel.

8.9 Write a short essay describing our current estimate of the fraction of stars that have planets, describing how we have made significant progress in this area over the last decade.

Chapter 9
Other Space Weather Phenomena

There is good evidence that within the last millennium the sun has been both considerably less active and probably more active than we have seen it in the last 250 years. These upheavals in solar behavior may have been accompanied by significant long-term changes in radiative output. And they were almost certainly accompanied by significant changes in the flow of atomic particles from the sun, with possible terrestrial effects.

(Eddy, 1976)

9.1 Key Concepts

- anthropogenic space weather
- asteroid impact
- global climate
- supernovas and gamma ray bursts

9.2 Learning Objectives

After actively reading this chapter, readers will be able to:

- describe other possible space weather impacts on Earth and arrange them in order of likelihood and consequence;
- compare and contrast natural and anthropogenic space weather effects.

9.3 Introduction

Can space weather have impacts on the **global climate**? We know that the amount of energy striking Earth from the Sun is the main driver of weather and climate. Temperature differences from day to night and from season to season are explained by the intensity of sunlight falling onto the surface and into Earth's atmosphere. For the most part, the change in intensity of sunlight is not due to changes in the luminosity or brightness of the Sun, but instead due to the daily rotation of Earth and the regular annual variation in hemispheric tilt toward or away from the Sun. In the northern hemisphere, summer is when Earth's northern hemisphere is tilted toward the Sun; winter in the northern hemisphere occurs six months later when Earth is on the opposite side of the Sun and the northern hemisphere is tilted away from the Sun.

We know that Earth's climate has changed often in the past. Most of these changes occurred gradually – over thousands, if not millions, of years. One regular climate cycle is called an Ice Age (with initial capital letters), when Earth moves from periods of global cooling and extensive ice cover to periods with less ice and and relatively warm temperatures. Ice Ages have come and gone regularly in Earth's history. Two to three million years ago we entered the most recent Ice Age – called the Quaternary. Ice Ages are marked by periods of extensive ice cover (ice ages with lower case letters) and interglacial periods with climate similar to today. The rise of agriculture and human civilization corresponds climatically to Earth leaving the last ice age (or entering an interglacial period) about 10 000 years ago. The present era is called the Holocene and is characterized by relatively mild global temperatures, with year-round ice cover restricted to high elevations and the polar regions. Some of these global climate changes correspond to changes in Earth's orbit (its eccentricity and its inclination) in cycles that last tens of thousands to hundreds of thousands of years, called Milankovic[1] cycles. Other large-scale changes have been brought about by plate tectonics that have changed the topology and landscape of the Earth – creating and destroy-ing seas, oceans, and mountains. In addition, life itself has changed Earth's climate by changing the chemical composition of the atmosphere. The first such large-scale change occurred soon after the arrival or formation of photosynthetic life, which released oxygen as a by-product of respiration. More recently, Earth over the last century has experienced warming that is unprecedented during the last thousand years and is due to the increase in greenhouse gases since the Industrial Revolution. Model predictions indicate that Earth's climate may be drastically different than it is today in the next few decades because of the burning of hydrocarbon fuel for transportation and energy. This is very quick compared to the normal timescale of climate change.

Does the Sun's luminosity or other aspects of the Sun's energy output change or is the 11-year solar cycle a permanent property of the Sun? We know that the Sun is constantly changing. These changes have been observed minute to minute, day to day, and year to year. Changes have also been observed over the centuries, the most dramatic of which is called the Maunder Minimum, when solar activity, as indicated by the number of sunspots, apparently disappeared. This

---

[1] Milutin Milankovic (1879–1958), Serbian engineer and mathematician who demon-strated that many periods of climate change correspond to changes in the orbital characteristics of Earth, such as its inclination and orbital eccentricity.

period coincided with a time of extremely cold winters and cool summers in Europe and North America now called the Little Ice Age. Is there a cause-and-effect link between solar variability and climate?

The Sun also isn't the only source of energy into Earth's space environment. Cosmic rays, asteroids, comets, and electromagnetic energy from other stars in our stellar neighborhood can also have an influence on Earth. Though events that can have significant impacts are very rare, they have happened in the relatively recent history of Earth. In addition, as we have learned, humans can change their climate; this is called anthropogenic climate change (meaning human-caused: "anthro" from the Greek for "human"). This is also true for space weather: the use of nuclear weapons can create **anthropogenic space weather** that can rapidly change the climate, creating artificial radiation belts and large induced currents. As the human population continues to grow, the impacts of some of these extreme natural and man-made space weather effects can be devastating. This chapter describes four space weather effects that could have significant and potentially catastrophic impact on civilization.

9.4 Climate Variability and Space Weather

How does changing solar activity, as indicated by the number of sunspots, affect our weather and climate? We know that the luminosity of the Sun hardly changes at all over the 11-year solar cycle (it is called the solar constant for a good reason). The total luminosity changes by about 0.1% between solar minima and solar maxima. Models of Earth's climate clearly show that changes at this level have a very small impact on global mean temperatures and large-scale weather patterns.

There are a considerable number of observations that show a significant correlation between long-term, decadal solar activity and climate, but it is not clear if the correlations signify direct causation, are due to random correlation, due to misidentification of similar frequencies because of other confounding events (such as volcanic eruptions), or due to uncertainties in the underlying data. One strong correlation that has been found in a climate proxy (a data set that is used to infer climate variability) is the amount of carbon 14 in tree rings. Carbon 14 is used in radio-carbon dating techniques to determine the age when a living organism died. Carbon 14 is incorporated into living organisms (including us) while we are alive and stops being accumulated when we or the plant dies. Using the radioactive half-life we can infer the date of death. Tree rings provide an annual record of growth, and the width and chemical and isotopic composition of the ring give clues on the environmental and atmospheric conditions during that year. Since some trees can live for centuries if not over 1000 years, they can provide an annual history of the climatic and environmental conditions of where they live. The variation of carbon 14 in tree ring data shows an anti-correlation with the solar cycle. This is explained by the fact that the primary formation mechanism of atmospheric carbon 14 is due to cosmic rays. Since cosmic ray flux is anti-correlated with sunspot number, there is an 11-year signal in tree ring data. Several anomalous carbon signals in the last 1500 years occurred during solar maximum periods and hence are thought to be proxies for major geomagnetic (Carrington-class) storms. However, this correlation demonstrates that a climate proxy (tree rings) can have variations with the same periodicity as the solar cycle that are not directly due to changes in climate, but to changes in cosmic ray intensity. So caution is warranted in inferring causation.

However, if it isn't the total amount of energy coming from the Sun, what could the physical mechanism be that connects climate and the Sun? With the advent of space-based observations of the Sun in the 1970s and 1980s, we have learned that the amount of solar UV radiation can change by 6–8% over a solar cycle. The amount of high-energy UV and X-ray emission from the Sun is correlated with the number of sunspots, so during solar maximum, solar activity and the amount of high-energy radiation from the Sun is at its peak.

High-energy EM radiation is absorbed in the thermosphere down to the stratosphere and influences the chemical reactions that can take place. Therefore changes in UV and X-ray intensity can have important consequences on the total energy balance of the atmosphere. The increase in high-energy solar radiation can directly heat the atmosphere and also cause warming indirectly through changing chemical reactions. We know that the thermosphere's temperature changes by a factor of 2 (from about 1000 K to 2000 K) over the solar cycle due to the enhanced high-energy solar radiation directly being absorbed. We also know that in the stratosphere increased UV leads to enhanced chemical reactions of oxygen atoms and molecules, increasing ozone formation. The increased ozone concentrations then absorb more UV, which leads to heating the equatorial stratosphere. This stratospheric warming then changes global wind patterns. In addition, increased solar activity increases the number of energetic particles that flow along magnetic field lines into the polar region, impacting the upper atmosphere. During the polar night, this particle precipitation leads to changed chemical reactions that can create large amounts of nitrogen and oxygen molecules (NO_x). Nitric oxide (NO) and nitrogen dioxide (NO_2) can circulate from the D region altitude mesosphere into the stratosphere where they lead to an increase in ozone destruction. Understanding the relative strength, spatial scales, and timescales of these two competing processes (increase of ozone and destruction of ozone) is one area of current active research.

So is solar activity – through the changes of high-energy UV and particle precipitation – another driver of climate change? Because of the relatively short history of our solar observations and accurate climate data, many of these correlations are very hard to confirm. Therefore a healthy amount of skepticism is present in the climate community. However, testable physical mechanisms are now being proposed linking space weather and climate, and new observations in the paleo-climate ("paleo" means ancient or pre-historic, from the Greek word meaning "long ago") indicate that there are correlations between a number of climate changes and solar activity changes. Some of the most recent dramatic climate changes, such as the warm period known as the Medieval Climatic Optimum that lasted from 900 to 1250, were associated with a period of intense solar activity. This period saw increased temperatures in at least the region around the North Atlantic. During this period the Norse people settled Greenland and gave it the name that now seems peculiar since it is covered by one of the world's largest ice sheets.

There are a number of causes of climate change, and Earth's climate has changed dramatically in the past. We are now coming to realize that our Sun is not a constant star and its variability can have a significant influence on Earth.

9.5 Asteroid and Comet Impacts

The Sun and cosmic rays are not the only energy inputs from space into Earth's atmosphere. Comets and asteroids have impacted Earth since its formation. One of the largest impacts with Earth occurred early in Earth's history when a large asteroid (about the same diameter as Mars, which is about half the size of Earth) hit Earth. Out of this collision, the Moon formed. Since then, Earth has been hit by asteroids and comets continuously, though the rate of impacts has dropped off considerably since about 3.8 billion years ago. The epoch of Earth from its formation 4.5 billion years ago to about 3.8 billion years is called the Hadean (from the Greek word "Hades," for "Hell"), because of the continuous bombardment of asteroids and comets that were present in the early solar system. Over the last 3.8 billion years, many of these objects have impacted with the planets and moons and each other, been swept out of the solar system, or confined to quasi-stable orbits such as the asteroid belt. However, there are still quite a few that cross Earth's orbit, and there is a small, but good, chance each year that one will hit Earth again. These impacts could have global climatic effects.

Some of the most rapid climate change events correspond to impacts of large asteroids with Earth. The most famous **asteroid impact** is the one that occurred 65 million years ago and is thought to have led to the end of the dinosaurs. Could such an impact happen again? The short answer is yes. Extraterrestrial dust, meteoroids, and asteroids continuously bombard Earth. Some make it to the surface of Earth. On any given day, about 100 tons of extraterrestrial material settles onto the surface of Earth. (Some of the dust on your windowsill might be interplanetary material.) Some of the dust or sand-sized objects "burn up" in the atmosphere as meteor trails or shooting stars. We know that a number of large objects cross Earth's orbit that have the potential to hit Earth at some point in the future. Objects that have diameters greater than 1 km can cause global effects. The number of near-Earth asteroids of this size is estimated to be about 800, and there is a probability of Earth being hit by one of these asteroids about once every million years. Asteroids smaller than this are much more likely to impact Earth, and though they probably wouldn't cause long-term global climate change, they can have detrimental and deadly local impacts. Plate 12 shows a picture of meteor crater in northern Arizona. It is about 20 000–50 000 years old and is 1200 m wide and 170 m deep. It was created by an iron–nickel meteor about 50 m in diameter. It is estimated that asteroids of this type and size hit Earth every few thousand years.

The most recent asteroid impact on the order of a diameter scale of 50–100 m was in 1908 when an asteroid slammed into the forests of Siberia near Tunguska. In 2013, a meteor of approximately 20 m diameter hit near Chelyabinsk, Russia (a city of over 1.1 million about 1800 km east of Moscow). The Chelyabinsk meteor was the largest meteor to hit Earth since the Tunguska event and is the only meteor impact known to have caused widespread injuries (1500 people sought medical attention; most injuries were due to flying glass that was blown out of windows by the shock wave as people inside buildings gathered at the window to watch the fireball streak across the sky). Asteroids less than 100 m in diameter generally don't hit the surface, but explode in the air (called an airburst). Fortunately the 1908 asteroid hit well away from populated areas, but it knocked down trees over a 25 km diameter area and released energy equivalent to an approximately 50 megaton explosion (the size of a large nuclear bomb). If this had hit a populated city, it would have been one of the most catastrophic natural disasters in history. The Chelyabinsk meteor was also an airburst at about 30 km altitude and exploded with an equivalent of about 500 kilotons of TNT. The atomic bomb dropped on Hiroshima had a yield of 150 kilotons. Fortunately much of the energy was absorbed by the atmosphere, but the blast wave that hit the ground was sufficient to cause building damage and injuries across six cities in the region.

When a large asteroid hits Earth, the atmospheric and climatic effect depends on whether it hits in the ocean or on land. Ocean impacts are more probable because oceans cover nearly three-quarters of Earth's surface. The impact of a large asteroid with the ocean would raise large amounts of water into the stratosphere and launch an incredibly large-scale tsunami. Some models predict that a mid-Atlantic Ocean impact of an asteroid of 1 km diameter would blast a hole in the ocean 11 miles across all the way to the seafloor and create a giant tsunami as water rushes back into the cavity. This tsunami could be a wave 100 m (400 feet) high when it hits the continental shelf off the coast of the USA. Obviously this wave would have impacts many kilometers inland and would devastate coastal communities. The effect of stratospheric water vapor could have short-term climate impacts similar to a large volcanic explosion. It is estimated that Earth has been hit about 600 times by a kilometer-sized asteroid since the time of the dinosaurs. Impacts by asteroids with diameters greater than 2 km would create a global impact that could directly kill several billion people.

Impacts on land can have global climate impacts because of the amount of dust ejected into the stratosphere and the potential for widespread fires. Stratospheric dust can have the effect of cooling Earth's surface temperature by blocking sunlight and changing the atmospheric composition. This cooling would be global and could last for years. Recent volcanic eruptions such as Mt. Pinatubo in 1991 had significant global temperature effects that lasted for several years. There is increasing geologic evidence that a period of rapid cooling (14 °F or 8 °C) occurred over just a few years about 12 800 years ago during what is called the Younger Dryas event. The lower temperatures lasted about 1400 years and corresponded to the decline of Pleistocene large animals (megafauna) like mastodons.

Currently we do not have a complete inventory of all the asteroids that could potentially impact Earth nor would we be capable of doing anything if we knew one was heading our way. NASA has a program called Near-Earth Objects (http://neo.jpl.nasa.gov/) that is attempting to identify all asteroids and comets that cross Earth's orbit and hence have a probability of hitting Earth. The relative yearly probability of being killed by an asteroid impact is very small, about 20 000 to 1 over a lifetime. However, the potential global impacts make these highly unlikely but very destructive events important in Earth's history. Figure 9.1 shows the average time between impacts as a function of the size of the impactor. Note that small impacts are highly probable over the scale of a human lifetime. A large asteroid impact, though

Figure 9.1 The frequency of asteroid impacts (or average time between impacts) versus size of impact in terms of kinetic energy released (joules along the bottom axis and megatons of TNT equivalent along the top axis). Examples of known impactors and the range of nuclear winter are labeled.

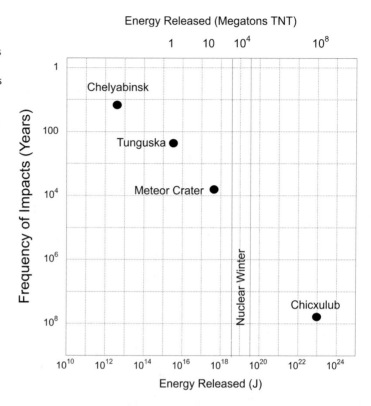

unlikely, is one of the most devastating natural hazards and has the potential to cause a mass-extinction event and destroy civilization.

9.6 Anthropogenic Space Weather

Humans have the ability to dramatically alter Earth and its environment. The burning of coal, natural gas, and oil is changing the composition of the atmosphere, leading to dramatic climatic changes that are impacting ecosystems and human society. Humans now move more dirt in mining, road building, and construction projects than all other natural process (erosion through rivers, wind, and other natural processes) and therefore are the major transformer of the Earth's surface. We have also developed nuclear weapons that can unleash huge amounts of energy with the potential to almost instantly change our global climate and kill hundreds of millions of people. The same general processes associated with large asteroid impacts that lead to prolonged cooling (increase of aerosols, smoke and dust blocking sunlight) from extensive fires and directly from the impact are the same processes that would happen during a nuclear explosion over major cities. Not only would significant numbers of people die from the initial explosion, those far from ground-zero would be impacted by the rapid temperature drops, which would last for years if not decades and would significantly impact agriculture. This scenario is called nuclear winter. In addition to these climatic effects, there are several other impacts that are essentially identical to those generated by a large coronal mass injection driving a geomagnetic storm – creating artificial radiation belts, increased ionization and structuring of the ionosphere, and the creation of large electric fields and currents, including large geomagnetic induced currents (GIC). These anthropogenic space weather events would destroy satellites, disrupt radio communication and navigation, and damage and destroy electronics and power systems.

The most devastating use of a nuclear weapon in terms of space weather impacts would be a high-altitude nuclear detonation – that is, detonating the bomb in the middle to upper atmosphere (25–75 km altitude) or in space (several hundred kilometers altitude). These explosions would generate a large electromagnetic pulse (EMP) that would create energetic particles and couple to the Earth's ionosphere and atmosphere, creating large electrical currents. These effects would extend for potentially hundreds to thousands of kilometers away from the blast, impacting large areas of the surface of the Earth. The radiation generating by the explosion and through secondary energization processes would create artificial radiation belts. The US Starfish nuclear test detonated a 1.4 megaton device at approximately 400 km altitude over the south Pacific in 1962. US spacecraft measured the creation of intense inner radiation belts and the failure of about one-third of operational satellites. In addition, the EMP created intense currents that caused power outages in Hawaii over 1400 km away.

The climatic impacts of even a limited nuclear exchange could cause global surface temperatures to drop below freezing and last for years, essentially creating years without a summer. The impact on agriculture would have far-reaching effects and lead to famine. A global nuclear war could drop temperatures to many tens of degrees below freezing for up to a decade or longer, destroying most ecosystems and agricultural productivity.

In a similar way to natural space weather, humans now have the capability to create our own space weather that would not only lead to myriad direct technological impacts (satellite destruction, radio and navigation disruption, and power grid failures), but could also create space weather impacts nearly identical to asteroid impacts, causing severe climate cooling.

9.7 Nearby Supernovas and Gamma Ray Bursts

There are processes in space that can create even more energy than asteroid impacts or nuclear explosions, but fortunately they occur very far away. However, because of the amount of energy they produce they can cause large space weather impacts on Earth that can be disruptive to catastrophic. Stellar explosions, called **supernovas**, send energy out into interstellar space in the form of electromagnetic and charged particle radiation that can impact the Earth's atmosphere, changing the energy and chemical balance.

There are two main types of supernova, Type I and Type II (astronomers are sometimes not very imaginative when it comes to naming things). Type I supernovas occur in binary star systems when a white dwarf star (the remnants of a dead Sun-like star) accretes or attracts material from its companion. The material can build up on the white dwarf star until the increase in pressure causes such an increase in internal temperature that the surface layer of the white dwarf star explodes. Type II supernovas occur when a star much more massive than the Sun exhausts its supply of nuclear fuel, thermonuclear reactions cease, and the star collapses under its own gravitational force. This collapse triggers a massive explosion that releases tremendous amounts of energy (about a billion billion times more energy than the Sun emits and generally several times more energy than a Type I supernova). This energy is in the form of electromagnetic radiation (gamma rays) and energetic particles that expand out into the universe (thought to be the source of galactic cosmic rays along with energetic particles that are accelerated by the expanding shock wave). What happens if there is a supernova in the neighborhood of Earth? If a supernova goes off within about 50–100 light-years (LY) of Earth, the energy input into Earth's space environment and atmosphere could be enough to dramatically change the photochemistry of the atmosphere and destroy the ozone layer. This would expose the surface to high doses of UV radiation from the supernova and the Sun. The effect could last for years, and the UV radiation dose could be 10 000 times normal levels. This would obviously have significant impacts on the biosphere. The size of the Milky Way galaxy is about 100 000 LY across, so a star within 100 LY is in our immediate neighborhood.

How many stars are within 100 LY of Earth? The estimate is about 14 000, though most of these are unknown since many stars are too dim to detect even within our stellar neighborhood. A vast majority of these stars are also not large enough to go supernova (most are red dwarf stars). How many of the remaining stars can go supernova? Currently there aren't any Type I or Type II candidate supernova progenitor stars within 100 LY (the closest Type I, IK Pegasi, is 150 LY away and the closest Type II, Spica, is about 250 LY away). However, in our galaxy, a star goes supernova about once a century. The last directly observed one was in 1604 and was recorded by Kepler. Cassiopeia A (over 11 000 LY away) is a supernova remnant that is dated to a supernova that occurred in 1680, but wasn't observed and recorded at the time; it took a few centuries for it to be identified with telescopic observations. More recently, a supernova remnant was observed in 1984, and subsequent observations showed that it was expanding rapidly. This allowed astronomers to date the initial supernova and they found that it occurred somewhere between 100 and 200 years ago (it is located near the galactic center about 25 000 LY away, so though the explosion seen from Earth occurred recently, the supernova explosion itself occurred long ago since it took the light 25 000 years to reach us). Of course our solar system and stars in the galaxy are all moving and therefore the distances between the Sun and other stars changes continuously. So over millions of years, stars that are far away from us can become closer. Taking data from all of the observed supernovas in the Milky Way along with estimates of how often stars that can go supernova are within 100 LY of Earth, scientists have obtained an estimate that the likely frequency of occurrence of supernovas in this region is about once every few hundred million years. The good news is that Earth is safe from a supernova explosion for at least a few million years. Therefore these very unlikely but highly destructive space weather events, which would have global catastrophic consequences, are fortunately not an immediate concern.

In addition to supernovas, astrophysicists often observe **gamma ray bursts** (or flares/flashes) from distant stars. These stars are called magnatars due to their extremely strong magnetic fields. They are collapsed neutron stars (stars that underwent gravitational collapse, but did not have enough mass to form a black hole). In 2004, one such gamma ray burst from a magnatar over 50 000 LY away ionized the atmosphere down to 20 km and essentially caused daytime ionization levels on the nightside. However, it lasted only a brief time and therefore did not impact technology significantly.

9.8 Supplements

9.8.1 Additional Learning Objectives

After actively reading these supplements, readers will be able to:

- calculate the kinetic energy of any moving object given its mass and velocity and be able to compare magnitudes of energy with respect to equivalent yield of atomic and nuclear weapons;

- evaluate the strengths and weaknesses of studies that use correlation of variables as part of the analysis, recognizing that without some physical mechanism connecting the variables, the significance may not be related to cause and effect.

9.8.2 Kinetic Energy and Conservation of Energy

Energy comes in many different forms: mechanical, chemical, and electrical energy, to name a few. Mechanical energy is the energy that an object has due to its motion (called kinetic energy) or its potential energy due to its stored energy of position. For example, a spring can contain potential energy if it is compressed or stretched. When the spring is released, the potential energy is converted to kinetic energy. Another example is holding a ball above the ground. The ball has gravitational potential energy due to its position above the ground. If dropped, the ball's potential energy will be converted to kinetic energy as it speeds up or accelerates to the ground.

Chemical energy is the potential energy contained in the bonds between atoms. When chemical reactions – such as the oxidation of molecular oxygen and methane gas – occur, energy is released as heat, light, and work (the gas expands). The amount of energy released depends on the type and amount of reactants in the chemical reaction. Some reactions release energy and are called exothermic ("exo" is Greek for "out of," while "thermic" is from the Greek "therme" for "heat"). Other reactions, which take energy from an external source and absorb energy into the product of the reaction, are called endothermic ("endo" is Greek for "within"). Biological systems use chemical energy to sustain life. The chemical reactions that sustain animal life involve the oxidation of sugars and other hydrocarbons. Humans also use chemical energy for transportation (burning of gasoline converts the chemical energy inside the hydro-carbon fuel into mechanical energy of the moving pistons to the driveshaft to the wheels) and environmental control (air conditioning and heating).

The concept of kinetic energy – the energy of motion – is important for understanding asteroid impacts since the amount of kinetic energy an asteroid has determines how destructive its impact would be on life on Earth. Kinetic energy is proportional to the mass of the object (the more mass a moving object has, the more kinetic energy it has) and the square of the velocity (the faster an object is moving, the more kinetic energy it contains). This is written as

$$KE = \frac{1}{2}mv^2.$$

The SI unit of energy is the joule. For very large energies, we can convert joules into kilotons of TNT (1 kiloton TNT = 4.184×10^{12} joules).

The conservation of energy states that the total amount of energy in a closed system must be conserved. In other words, you can convert the system's energy from one form to another (from potential to kinetic, or chemical to mechanical), but you cannot get more energy out of the system than it already contains without adding it. This principle allows easy calculation of a number of important variables regarding the motion of the object. For example, one may wonder how fast a one-kilogram mass is moving right before it hits the floor after being dropped from a height of 1 m. We know from the conservation of energy that the potential energy of the system can be converted to kinetic energy. One way to express the potential energy of a mass lifted above the ground is

$$PE = mgh,$$

where PE is potential energy, m is mass, g is the acceleration of gravity at Earth's surface, and h is the height above the ground. This can be set equal to the kinetic energy of the object to find the object's velocity just before it hits the ground since the potential energy will be converted to kinetic energy. Therefore,

$$PE = mgh = KE = \frac{1}{2}mv^2,$$

$$v^2 = 2gh.$$

The one-kilogram mass will be moving 4.4 m s^{-1} just before it hits the floor. Note that the mass of the object actually cancels out of the equation. The velocity of an object dropped from a height is independent of the object's mass and depends only on the acceleration of gravity and the height from which it is dropped.

For the asteroid problem, we can estimate the velocity and mass of the object and therefore can estimate the total amount of kinetic energy that will be released. We use the conservation of energy to understand into what forms of energy this kinetic energy will be converted (i.e., heat, sound, the movement of dirt, rock or water at the impact site, the heat of vaporization, etc.). For asteroids more than 2 km across, the amount of energy released would be globally catastrophic. Even asteroids less than 2 km in size can have major regional impact and could directly affect millions of people, causing untold deaths and hundreds of billions of dollars in economic damage (see Chesley and Ward, 2006, for a description of the probabilities and damage estimates of various sizes of asteroids and their human and economic impacts).

9.8.3 Correlation and Causation

One of the goals of science is to understand how and why things work. Knowing the cause of an effect can help us predict a future outcome. (For example, we know that cigarette smoking causes lung cancer and a number of other illnesses. Therefore, if we prevent young people from smoking, we can reduce the overall health costs and impacts of a number of illnesses.) One way to attempt to find the cause of a phenomenon is to study a number of other parameters and see whether any are correlated with the phenomenon of interest. For example, scientists studying hurricanes examined the role of sea-surface temperature, atmospheric pressure, and wind velocities as a function of altitude. Meteorologists found that warm sea-surface temperatures can give rise to the formation of tropical storms and hurricanes, and that upper-level winds can be detrimental to the development of a storm. With this information, forecasters and meteorologists can make physical models in order to predict the development and evolution of a hurricane. Of course, scientists must make some assumptions or initial hypotheses about what parameters to examine when looking for correlations. This is done by thinking of possible physical mechanisms that would relate one parameter to a certain outcome or phenomenon. For example, in space weather, one may suspect that the direction of the interplanetary magnetic field would have an impact on the magnitude of a geomagnetic storm. This follows from our understanding of magnetic reconnection – southward IMF can directly couple with Earth's northward-directed magnetic field and transfer energy from the solar wind into Earth's magnetosphere. The excellent correlation between southward IMF and geomagnetic activity provided a strong piece of evidence that magnetic reconnection plays an important role at the dayside magnetosphere. Of course, scientists could have examined a number of things to see if they are correlated with geomagnetic activity. These include the price of gold, the performance of a sports team, the crime rate, the phase of the Moon, or literally an infinite number of other things. What may be pretty surprising is that if you looked at enough random variables, there is a good chance that some of them *would* be correlated with geomagnetic activity (or whatever else you are examining). Most things tend to have periodicities or cyclic behavior. And sometimes, by chance, two phenomena can appear to be correlated.

Another complication is that one phenomenon can be correlated with another and not be the direct cause even if there may be a physical mechanism that potentially connects the two phenomena. This is because many factors may be correlated with each other even though they are independent. For example, there is a strong correlation between illiteracy and high mortality due to a number of diseases. There isn't a direct causative link between deadly disease and being able to read, but there is a clear and strong correlation between these two phenomena. The solution to this puzzle is that there is a strong correlation between low socioeconomic status and illiteracy. There is also a strong correlation between low socioeconomic status and access to healthcare and good nutrition. Therefore illiteracy isn't the direct cause of deadly diseases, and neither is being poor. The cause is lack of access to healthcare and good nutrition.

Therefore, critical thinkers are careful when attempting to find the cause of an effect not to confuse correlation and causation. One must always have a clear physical mechanism in mind when linking two variables, and one must always test whether a correlation is pure coincidence or actually related.

9.9 Problems

9.1 Some recent studies suggest that the strength of the solar magnetic field is decreasing. What would the potential climate impact be of a prolonged period of weak solar magnetic field?

9.2 What role does the ozone layer play in protecting life on Earth's surface? What role does Earth's magnetosphere play?

9.3 Kinetic energy is the energy of motion. How much kinetic energy does a spherical asteroid with a 1 km diameter have if it is made out of iron (iron has a density of about 8 kg m^{-3}) and hits Earth at a relative velocity of 10 km s^{-1}? Convert your answer to megatons of TNT (1 megaton of TNT = 4.18×10^{15} joules).

9.4 Should society be concerned about low-probability events such as asteroid impacts? We know that there are many natural drivers of climate change. Should society be concerned about global climate change that can take decades to manifest itself?

9.5 Would there be any warning of a nearby supernova? Why or why not?

9.6 If rapid global impacts (and consequent mass-extinction events) have happened in the past every few hundred million to billion years due to asteroid impacts and supernova, how can we explain our existence today?

9.7 How does a Carrington-class event fit within the threats to society from other space weather hazards? Make a concept map of different cataclysmic space weather impacts and identify similarities and differences.

9.8 One can find decadal variability in a wide range of human activity (from health to economic) and even find direct correlation between space weather effects (geomagnetic storms, cosmic ray intensity) and these human impacts. Explain why a strong correlation may or may not connect two phenomena. (See www.tylervigen.com/spurious-correlations to discover other correlations.)

Appendix 1
Basic Math Rules

Math follows rules using a set of symbols to determine the result of an equation. There are a set of basic operators (addition, subtraction, multiplication, division) that are taught early in most educational systems. The rules (like the multiplication table) are memorized so you can "do math" easily and quickly. Calculators, spreadsheets, and computer programming have enabled much of math to be done by a machine. However, there is utility in being able to do simple arithmetic and algebra to help quantitative understanding of many problems. Below are some basic math rules beyond simple arithmetic that will aid your quantitative reasoning.

Exponent Rules

Rule 1: When multiplying numbers with exponents, simply add the exponents.

$$x^m \times x^n = x^{m+n}.$$

Example

$$10^4 \times 10^5 = 10^{4+5} = 10^9.$$

Rule 2: When dividing numbers with exponents, simply subtract the exponents.

$$x^m \div x^n = x^{m-n}.$$

Example

$$10^7 \div 10^5 = 10^{7-5} = 10^2.$$

Rule 3: When taking a power of a number with an exponent, simply multiply the exponents.

$$(x^m)^n = x^{mn}.$$

Example

$$(10^2)^4 = 10^{2\times4} = 10^8.$$

Rule 4: The power of a product is simply the product of each parameter raised to the exponent.

$$(xy)^m = x^m \times y^m.$$

Example

$$(4 \times 5)^2 = 4^2 \times 5^2$$
$$(4 \times 5)^2 = 16 \times 25$$
$$(4 \times 5)^2 = 400.$$

Rule 5: The power of a fraction is simply the quotient of each parameter raised to the exponent.

$$(x/y)^m = x^m/y^m.$$

Example

$$(3/4)^2 = 3^2/4^2$$
$$(3/4)^2 = 9/16.$$

Rule 6: Negative exponents signify to take the reciprocal.

$$x^{-m} = 1/x^m.$$

Example

$$2^{-2} = 1/2^2$$
$$2^{-2} = 1/4.$$

Rule 7: Any number raised to the power zero (0) is equal to 1.

$$x^0 = 1.$$

Example

$$5^0 = 1.$$

Rule 8: Any number raised to the power of 1 is equal to the number.

$$x^1 = x.$$

Example

$$5^1 = 5.$$

Rule 9: You can algebraically solve for a parameter that has an exponent by raising both sides by the reciprocal.

$$x^{m/n} = y \rightarrow x = y^{n/m}.$$

Example

$$x^{1/2} = 3$$
$$x = 3^{2/1}$$
$$x = 3^2$$
$$x = 9.$$

Rule 10: A fraction raised to a negative exponent is the reciprocal raised to the exponent.

$$(x/y)^{-m} = (y/x)^m.$$

Example

$$(2/3)^{-2} = (3/2)^2$$
$$(2/3)^{-2} = 3^2/2^2$$
$$(2/3)^{-2} = 9/4.$$

Order of Operations

Some math may have multiple parts to the equation (e.g., multiplication, division, addition, and subtraction in one equation). To work on problems such as these you need to know the correct order in which to do the operations. The rules for order of operation state which parts of the equation to do first, regardless of the order of the math written left to right. In the USA, the acronym PEMDAS describes the order of operation: P (parentheses), E (exponents), M (multiplication), D (division), A (addition), and S (subtraction). In the UK and India, the acronym BODMAS (or BEDMAS) is used, where B stands for "brackets" and O for "orders" (powers/indices or roots).

Of course, there are some special rules when you have both multiplication and division or both addition and subtraction in one equation. You should do the order of operation from left to right (e.g., M does not always come before D, and A does not always come before S).

Examples

$$25 \div 5 \times 3 = 5 \times 3 = 15$$
$$24 - 9 + 7 = 15 + 7 = 22.$$

Note that if you do the order of operation M first then D, or A first then S, instead of from left to right, the answers are different, so order of operation matters.

Appendix 2
SI Units

Scientists and engineers all over the world use a special set of units called the Système International d'unités (SI), which allows them to communicate quantitative information consistently. The fundamental SI units and derived SI units are shown in Tables A.2.1 and A.2.2.

Table A.2.1 *Fundamental SI units (see http://physics.nist.gov for more information about SI units)*

Base quantity	Name	Symbol
length	meter	m
mass	kilogram	kg
time	second	s
electric current	ampere	A
thermodynamic temperature	kelvin	K
amount of substance	mole	mol
luminous intensity	candela	cd

Table A.2.2 *Derived SI units*

Derived quantity	Name	Symbol
area	square meter	m^2
volume	cubic meter	m^3
speed, velocity	meter per second	$m\,s^{-1}$
acceleration	meter per second squared	$m\,s^{-2}$
mass density	kilogram per cubic meter	$kg\,m^{-3}$
current density	ampere per square meter	$A\,m^{-2}$
magnetic field strength	ampere per meter	$A\,m^{-1}$
luminance	candela per square meter	$cd\,m^{-2}$

For ease of understanding and convenience, several SI derived units have been given special names and symbols. A selection of those used in space weather studies are given in Table A.2.3.

Table A.2.3 *SI-derived units with special names and symbols*

Derived quantity	Name	Symbol
frequency	hertz	Hz
force	newton	N
pressure, stress	pascal	Pa
energy, work, quantity of heat	joule	J
power, radiant flux	watt	W
electric charge, quantity of electricity	coulomb	C
electric potential difference, electromotive force	volt	V
capacitance	farad	F
electric resistance	ohm	Ω
electric conductance	siemens	S
magnetic flux	weber	Wb
magnetic flux density	tesla	T
inductance	henry	H
Celsius temperature	degree Celsius	°C
luminous flux	lumen	lm
activity (of a radionuclide)	becquerel	Bq
absorbed dose, specific energy (imparted), kerma[1]	gray	Gy
dose equivalent (d)	sievert	Sv

[1] Kinetic energy released per unit mass ($J\,kg^{-1}$ or gray)

Appendix 3
Useful Space Physics Formulary

Alfvén velocity $= \dfrac{B}{\sqrt{\mu_0 \rho}}$, where B is the magnetic field, μ_0 is the
permeability of free space, and ρ is the mass density.

Atmospheric drag $F(\text{air drag}) = A C_d \rho V^2$, where F is force, A is cross-sectional area, C_d is the drag coefficient, ρ is the mass density, and V is velocity.

Doppler shift $f' = f_0 \left(\dfrac{v \pm v_0}{v \pm v_s} \right)$, where f' is the perceived frequency; f_0 is
the actual frequency; v is the speed of the waves; v_s is the speed of the source and added to the wave speed if the source is moving away from the observer and subtracted if the source is move toward the observer; and v_0 is the speed of the observer and is added to the speed of the wave if moving towards the source and subtracted in moving away.

Electromagnetic photon energy $= hf$, where h is Planck's constant and f is frequency.

Gas pressure $= nkT$, where n is number density, k is Boltzmann's constant, and T is temperature.

Lorentz force $\vec{F} = q\vec{v} \times \vec{B}$, where q is charge, v is velocity, and B is magnetic field. The arrows over the symbols denote that they are vectors.

Magnetic pressure $= \dfrac{B^2}{2\mu_0}$, where B is magnetic field and μ_0 is the
permeability of free space.

Plasma frequency $= \omega_p = \sqrt{\dfrac{n_e e^2}{m \varepsilon_0}} = 9\sqrt{n_e}$ (SI units), where n_e is electron number density, e is electron charge, m is mass, and ε_0 is the permittivity of free space.

Plasma ram pressure $= \rho v^2$, where ρ is the mass density and v is velocity.

Wien's law $\lambda_{\text{peak}} T = 2.898 \times 10^{-3}$ m K, where λ_{peak} is the peak wavelength and T is temperature.

Appendix 4
Space Weather Timeline

2134 BC	The first mention of an ancient solar eclipse is made by Chinese astrologers Hsi and Ho on October 22, who were beheaded for not predicting its arrival accurately.
800 BC	First record of naked-eye observations of sunspots is made by Chinese astronomers.
600 BC	Discovery of lodestone and magnetic forces is cited in an ancient Chinese story of a navigation device that worked in a fog and kept pointing "south."
AD 37	First of many records of crimson lights in the northern sky is recorded in the story of Julius Caesar, who has his army march to a northern town in Italy to investigate its burning, which lit up the sky from Rome. The term "aurora" was later coined in 1621 by the French astronomer Pierre Gassendi (1592–1665).
1527	First definite printing of auroral descriptions appear in three separate pamphlets.
1600	William Gilbert's work *De magnete* is published, in which the Earth's magnetic field is first described.
1613	In his *Letters on Sunspots*, Galileo demonstrates that sunspots are at the solar surface and are not little "planets."
1645–1715	Sunspots are almost completely absent during this Maunder Minimum period, and extreme winters are experienced throughout the period. Potential links between space weather and terrestrial weather are controversial and continue to be a topic of ongoing research.
1716	Edmund Halley notes that auroral curtains are aligned with projections of the Earth's magnetic field into the upper atmosphere.
1740	Anders Celsius (1701–1744) and Olof Hiorter (1696–1750) discover that compass needles move in an erratic

manner instead of always pointing north. These "magnetic storms" were later investigated in more detail by Alexander von Humboldt (1769–1859).

1814   Joseph von Fraunhofer (1787–1826) reveals the solar spectrum using a new instrument called the spectroscope. By passing sunlight through a slit and a prism, he found faint black lines speckled the "rainbow," each a fingerprint of a specific atom. These elements were later identified by Gustav Kirchhoff (1824–1887) and Anders Angström (1814–1874) in 1863.

1832   Gauss develops the first mathematical description of the geomagnetic field.

1834   Michael Faraday, who discovered that magnetic fields can induce electric currents, first coins the word "ion."

1834   Baron von Humboldt proposes the first worldwide array of magnetometers to study geomagnetism.

1843   Heinrich Schwabe (1789–1875) uses the accumulated records of sunspots to uncover the 11-year sunspot cycle.

1852   Col. Edward Sabine (1788–1883) demonstrates that geomagnetic variations are a worldwide phenomenon using British observatories set up at the suggestion of Humboldt. He announces that the Sun's 11-year sunspot cycle is "absolutely identical" to Earth's 11-year geomagnetic cycle.

1852   Rudolf Wolf independently hypothesizes a link between the solar cycle and geomagnetic activity shortly after Sabine.

1859   Richard Carrington (1826–1865) and Robert Hodgson (b. 1804) first report a solar flare during a major week-long solar "superstorm" in 1859 that caused worldwide telegraph outages and brilliant sky-filling aurorae – they infer a causal link between the Sun and magnetic storms.

1864   James Clerk Maxwell (1831–1879) is the first to combine all four equations of electromagnetism in a unified framework, called Maxwell's equations, which have played a central role in modern physics by contributing crucially to contemporary field theory, relativity theory, and quantum theory.

1871   Karl Hornstein demonstrates that terrestrial magnetism exhibits a 26 1/3-day periodic change, which he suggests is related to the Sun's 27-day rotation period.

1871	Becquerel suggests that particles from the Sun are responsible for the aurorae; this inference is enhanced by George Francis FitzGerald in 1892, and by Kristian Birkeland and Sir Oliver Lodge in 1900.
1881–1882	Modern auroral physics begins with the first international cooperative study of polar regions, the First Polar Year. Sophus Tromholt discovers the auroral oval in 1881 and his First Polar Year study report of 1885 contains excellent auroral descriptions.
1892	Based on late nineteenth-century physics theory, Lord Kelvin states with "absolutely conclusive evidence" that "the supposed connexion between magnetic storms and Sun-spots is unreal, and that the seeming agreement between periods has been mere coincidence."
1892	Astronomer George Ellery Hale (1868–1938) creates a camera that views the Sun through light produced only by hydrogen atoms and produces the first "hydrogen-alpha" images of solar prominences and flares.
1896	Kristian Birkeland proposes that aurorae and magnetic storms are caused by beams of very fast electrons emitted by the Sun.
1900–1903	Based on expeditions in 1899 through 1903, Kristian Birkeland proposes that the same particles that produce aurorae also cause geomagnetic variations, which he discovered and named polar elementary storms, now called substorms.
1902	Oliver Heaviside suggests that a conducting layer in the upper atmosphere allows radio waves to propagate around the Earth's curvature; this Heaviside layer is now called the ionosphere.
1904	E. Walter Maunder confirms the earlier results of Hornstein in 1871 and Airy in 1872 that linked the 27-day recurrence of geomagnetic activity with the rotation rate of the Sun.
1907	Carl Størmer develops the first quantitative theory of charged particle motion in Earth's magnetic field.
1908	Kristian Birkeland suggests a cavity around Earth for solar corpuscles, and infers a ring current and continuous solar wind, all confirmed in future decades.
1908	George Ellery Hale discovered solar magnetism and suggested that sunspots were not sufficient to create storms and the origin of storms "may be sought with

	more hope of success in the eruptions shown … in the regions surrounding spots."
1910	Birkeland carries out experiments with his terrella apparatus, which simulates Earth's magnetosphere through interactions of electrons beams and a magnetized sphere within a vacuum chamber.
1919	Lindemann suggests solar particle streams of quasi-neutral ionized "gas" – later called plasmas.
1923	Based on experimental work with electron and positive ion systems, Irving Langmuir coins the term "plasma."
1923	Sydney Chapman develops a theory of magnetic storms using Lindemann's suggestion of 1919.
1924	Sir Edward Appleton uses radio waves to demonstrate the existence of the ionosphere and infers the existence of more than one ionized layer in that region.
1931	Appleton develops a theoretical basis for ionospheric structure.
1939	Hannes Alfvén proposes that the Earth is imbedded in a general electric field from the solar wind; he provides field theory for magnetic storms and aurorae.
1942	Alfvén develops powerful approximation for plasma dynamics called magnetohydrodynamics (MHD), leading much later to the 1970 Nobel Prize.
1946	Ronald Giovanelli makes the first reference to the reconnection process.
1950	Hannes Alfvén creates an MHD theory in which aurorae are treated as an electric discharge.
1951	Ludwig Biermann predicts the presence of solar wind based on the analysis of comet tails.
1953	Owen Storey uses radio dispersion from lightning to infer the existence of plasmas in space beyond the ionosphere.
1955	Leverett Davis, Jr. suggests the existence of the "heliosphere," which is the region of space formed by the interaction of interstellar plasmas with solar plasmas and the interplanetary field.
1956	Physicist Philip Morrison (1915–2005) explains why it is that the energies of cosmic rays follow a particular pattern, and proposes that Earth is surrounded by a vast cavity of plasma over 200 AU in diameter, later known as the heliosphere.
1956	Physicist Peter Sweet proposes the first magnetic reconnection model for solar flares, later called the Sweet–Parker theory of flares.

1958	Eugene Parker develops the MHD plasma model for the solar wind and interplanetary magnetic fields.
1958	The first satellite launched successfully into space by the USA carries a Geiger counter that registers countless radiation events. Physicist James Van Allen (1914–2006) concludes that Earth is surrounded by belts of high-energy particles, the radiation belts, also named the Van Allen belts in his honor.
1959	Thomas Gold coins the word "magnetosphere."
1960	Paul Coleman makes the first in situ observations of the interplanetary magnetic field.
1960	Physicist Robert Leighton borrows ideas from seismology to propose a new method to study the interior of the Sun, called helioseismology; however, this technique is not applied until the SOHO satellite in the late 1990s.
1961	James Dungey develops an open model of the magnetosphere which incorporates dayside and nightside reconnections to explain the coupling between the solar wind and the magnetosphere, now called the Dungey cycle.
1962	Although a plasma flow from the Sun was inferred by Biermann in 1951 based on analyses of comet tails, Marcia Neugebauer is the first to directly observe this "solar wind" using data from the Mariner 2 spacecraft en route to Venus.
1962	L. J. Cahill and P. G. Amazeen discover the magnetopause through direct observations of Earth's outer magnetospheric boundary from Explorer 12; Charles Sonett provides the first observations of an interplanetary shock (1962), and Norman Ness confirms Earth's bow shock (1963).
1963	Physicist Donald Carpenter proposes that Earth's atmosphere extends thousands of kilometers into space in a region we now call the plasmasphere. Apollo missions between 1968 and 1972 photograph the neutral component of this region, which is called the geocorona.
1964	Syun-Ichi Akasofu connects auroral morphological changes to geomagnetic disturbances and introduces the concept of geomagnetic substorms.
1965	Norman Ness discovers the Earth's magnetotail.
1966	Charles Kennel and Harry Petschek formulate a quantitative theory of trapped radiation in Earth's magnetosphere, including plasma physics limits on stably trapped particle fluxes.

1966	Sam Bames Jr. discovers Earth's "plasma sheet," a major component of the magnetotail.
1966	Don Fairfield discovers that the level of geomagnetic activity depends on the polarity of the interplanetary magnetic field (IMF).
1966	John Spreiter develops first quantitative model of solar wind flow around the magnetosphere.
1971	Discovery by OSO-7 of bursts of plasma from the solar surface, which were called coronal transients and later named coronal mass ejections (CMEs). The first drawing of a CME appeared in solar eclipse sketches of the corona in 1860.
1972	Ed Shelley discovers heavy ions (oxygen) in Earth's magnetosphere.
1972	Ed Hones discovers the magnetotail boundary layer based on Vela magnetometer data, precursor for the discovery of the plasma mantle (1974) and the Low-Latitude Boundary Layer (LLBL) (1976).
1973	Early rocket studies after World War II showed that the Sun emits X-rays, which cannot penetrate Earth's lower atmosphere. The first actual images of the X-ray Sun are made by Skylab and show dark "coronal holes."
1974	Don Gurnett, a pioneer in radio wave observations of Earth's magnetosphere, discovers that Earth is a significant radio source.
1976	Takesi Iijima and Thomas Potemra develop the first global map of field-aligned currents entering the polar regions of Earth's magnetosphere.
1979	Goetz Paschmann provides the first in situ detection of reconnection, observed by the ISEE spacecraft at Earth's magnetopause.
1979	The first detection of aurorae on another planet is made by the Pioneer 11 spacecraft as it flies by Saturn.
1979	The plasma sheet boundary layer of Earth's magnetotail is discovered by R. J. DeCoster and L. A. Frank using IMP 7 data.
1989	A large geomagnetic storm knocks out the Hydro Quebec power grid and several communication satellites in one of the largest space weather events during the modern age.
1992	The Hubble Space Telescope provides the first photograph of aurorae in the polar regions of Jupiter.

1995	The launch of the Solar Heliophysics Observatory (SOHO) ushers in the era of 24/7 solar observations, enabling new discoveries about coronal mass ejections (CMEs) and highlighting the differences between solar flares and CMEs outlined by Jack Gosling.
1997	In the same way that Earth has atmospheric and oceanic currents flowing steadily across thousands of kilometers, the SOHO satellite discovers the same kinds of "jet streams" within the 10 000 K plasmas at and below the solar surface.
2001	The launch of the TIMED (Thermosphere • Ionosphere • Mesosphere • Energetics and Dynamics) mission ushers in the long-running (still operating in 2022) investigation of the Earth's upper atmosphere, including understanding the impact of solar flares on the ionosphere and thermosphere.
2003	More than 45 years after the discovery of Earth's radiation belts, the IMAGE satellite discovers how ordinary lightning on Earth can change the shape of these Van Allen belts.
2006	The launch of the Solar Terrestrial Relations Observatory (STEREO) provides the first stereographic images of a CME as a pair of spacecraft orbit the Sun ahead of and behind the Earth.
2007	The launch of Vassilis Angelopoulos' Time History of Events and Macroscale Interactions during Substorms (THEMIS) mission with its phased-orbit constellation of five spacecraft enables the determination of the role of magnetic reconnection in the development of auroral substorms.
2012	Voyager 1 passes through the Sun's heliopause crossing into interstellar space at a distance of 122 AU (18 billion km), becoming the first spacecraft to leave the heliosphere. Voyager 2 enters the interplanetary medium in 2018.
2012	The launch of the twin spacecraft Van Allen Probes enable one of the most comprehensive investigations of processes responsible for the acceleration and transport of the Earth's radiation belts, named for James Van Allen, whose 1958 observations from the first US satellite discovered them.
2012	Dave McComas' interpretation of observations from the Interstellar Boundary Explorer (IBEX) mission find

that there is a bow wave instead of a bow shock upstream of the Sun's heliosphere. This is due to the relative speed of the Sun with respect to the interstellar medium being subsonic.

2018   The launch of the Parker Solar Probe mission heralds the first observations of the solar corona near the Sun (well inside Mercury's orbit) and provides insights into the processes responsible for heating the outer atmosphere of the Sun. Named in honor of Eugene Parker, who first described the theory of the solar wind and interplanetary magnetic field in 1958.

2021   Tom Immel presents the first direct evidence of the long-theorized wind-driven electric dynamo in the Earth's upper atmosphere using data from the Ionospheric Connection Explorer (ICON) spacecraft.

2022   Starlink loses 40 of 49 satellites due to increased atmospheric drag induced by a relatively minor geomagnetic storm soon after their launch. This is the largest number of satellites lost due to a single space weather event.

Appendix 5
Web Resources

There are many excellent websites developed that relate to space weather. Below is a selection of sites that are useful for exploring different aspects of the Sun–Earth relationship.

A regularly updated page giving news and information about the Sun–Earth environment:

www.spaceweather.com

The Space Environment Center continually monitors and forecasts Earth's space environment; provides accurate, reliable, and useful solar–terrestrial information; conducts and leads research and development programs to understand the environment and to improve services; advises policy makers and planners; plays a leadership role in the space weather community; and fosters a space weather services industry. The Space Weather Prediction Center is the USA's official source of space weather alerts and warnings:

www.swpc.noaa.gov/

A listing of satellite outages and failures:

www.sat-index.com/failures/

Space Weather: The International Journal of Research and Applications is an online publication devoted to the emerging field of space weather and its impact on technical systems, including telecommunications, electric power, and satellite navigation:

https://agupubs.onlinelibrary.wiley.com/journal/15427390

The European *Journal of Space Weather and Space Climate*, a link between all the communities involved in space weather and in space climate, including (but not limited to) space, solar, atmospheric scientists, engineers, forecasters, social scientists, economists, physicians, insurance experts, etc ·

www.swsc-journal.org/

NASA's Heliophysics webpage providing information on the missions and science being pursued by NASA:

https://science.nasa.gov/heliophysics

The High Altitude Observatory (HAO) explores the Sun and its effects on the Earth's atmosphere and physical environment, in partnerships extending throughout the national and international scientific communities for research, observational facilities, community data services, and education:

www2.hao.ucar.edu/

The European Space Agency's space weather web server:

https://swe.ssa.esa.int/current-space-weather

The National Solar Observatory webpage provides images and data of the Sun from a number of solar observatories:

https://nso.edu/

NASA's Coordinated Data Analysis Web provides access to all NASA data sets and tools to visualize and analyze the data:

https://cdaweb.gsfc.nasa.gov/

NASA's Community Coordinated Modeling Center provides access to community space weather models:

https://ccmc.gsfc.nasa.gov/

The UK Met Office's Space Weather page:

www.metoffice.gov.uk/weather/specialist-forecasts/space-weather

News site archiving stories about space weather:

www.space.com/topics/space-weather

"To understand a science it is necessary to know its history" (Auguste Comte). A timeline of the major achievements in our understanding of the solar–terrestrial relationship:

https://space.engin.umich.edu/outreach/timeline-of-space-physics/

The author's (Mark Moldwin's) University of Michigan webpage:

https://space.engin.umich.edu/

Glossary

Some useful space weather terminology is defined here.

anthropogenic space weather describes impacts due to nuclear weapons detonated in the atmosphere or space that mimic the natural space weather impacts such as geomagnetic induced currents (GIC) and enhanced radiation belts.

asteroid impacts are events when a rock from outer space hits Earth. The larger the asteroid the more energy it carries, creating larger effects on Earth. Asteroids are one of the potential global threats to civilization due to their large possible climatic effects.

atmospheric drag describes the frictional force or acceleration felt by an orbiting object due to its relative velocity with the gases in the upper atmosphere. For satellites it causes a loss of altitude that brings the satellite into a higher-density region that can ultimately destroy the satellite.

aurorae are emission of photons of light from upper atmospheric gases that are excited by impacts with charged particles traveling along magnetic field lines into the upper atmosphere.

Carrington-class event is a rare, extremely severe space weather storm comparable to the September 1859 event described by Richard Carrington. It is often used to describe worst-case space weather storms.

chromosphere is a layer of the solar atmosphere located just above the photosphere and is visible during total solar eclipses.

corona is the outermost region of the solar atmosphere characterized by temperatures of a million kelvin.

coronal mass ejections are large solar eruptions that carry part of the solar corona (about 10^{12} kg of magnetized plasma) outward at high velocities (up to 1000 km s^{-1}). They are the major cause of space weather effects on Earth.

corpuscular (or particle) radiation consists of the charged nucleus of an atom or an electron that with sufficient energy can ionize matter. The solar wind, galactic cosmic rays, solar energetic particles, and radiation trapped in the Van Allen radiation belts all have space weather impacts.

cosmic rays are energetic charged nuclei and electrons that are created in very energetic processes on the Sun, and by interplanetary shocks (called solar energetic particles), or from outside the solar system (called galactic cosmic rays), created by very energetic processes such as supernova and other stellar interactions. The "ray" is a misnomer as they are corpuscular (particle) radiation.

Dungey cycle describes the dynamics and convection of the magnetosphere due to dayside and nightside magnetic reconnection.

electromagnetic radiation carries energy and is classically described as electric and magnetic waves that propagate at the speed of light (c) at a given wavelength and frequency. The frequency of the radiation (from low-frequency radio waves to high-frequency X-rays and gamma rays) determines the energy of the radiation, with energy increasing with higher frequency. It can also behave like a particle called a photon.

exoplanets are planets orbiting other stars. We have discovered more than 5000 exoplanets as of June 2022, some of which are similar in size to Earth and located in the habitable zone (at the right distance from the parent star for liquid water to exist on the surface).

Faraday's law of induction describes how changing magnetic fields can induce a current in a circuit and is the basis for electric generators and geomagnetic induced currents (GIC).

geomagnetic induced currents (GIC) are transient currents induced by changes in the geomagnetic field driven by space weather that can flow through ground-level technical systems such as power transmission lines, communication lines, and pipelines, potentially damaging the systems.

geomagnetic storm is defined by the intensification of the ring current and is one of the dominant modes of response of the magnetosphere to energy transfer from the solar wind.

geosynchronous (or geostationary) orbit is a circular orbit above the geographic equator at approximately 6.6 r_E (Earth radii) that has an orbital period of one day (24 hours) and therefore appears "stationary" with respect to the ground since it orbits around the Earth at the same rate of the Earth's rotation. Used primarily for Earth observing and communication satellites.

global climate describes the average meteorological and weather conditions over the whole Earth. As humans continue to burn fossil fuels releasing carbon into the atmosphere, the Earth system changes (atmospheric and oceanic temperatures rise, ice melts, there is increased extreme weather), which leads to ecosystem and human system disruptions.

heat transfer is the transfer of energy through convection (fluid motion), conduction (absence of fluid motion such as through contact of solid to solid), or electromagnetic radiation.

heliosphere is the region surrounding the Sun that is defined by the solar wind and interplanetary magnetic field (IMF). The heliopause is the boundary between the heliosphere and the local interstellar medium.

high Earth orbit (HEO) has apogee beyond geosynchronous orbit (GEO). HEO is sometimes used to describe highly elliptical orbits that have apogees outside of medium Earth orbit and perigees in low Earth orbit (LEO).

interplanetary magnetic field (IMF) is the Sun's magnetic field that is drawn out by the radial motion of the solar wind and nominally has an Archimedean or Parker spiral configuration due to the rotation of the Sun.

interstellar medium (ISM) consists of the neutral gas, plasma, and dust in the region between the stars.

ionization is the process of converting a neutral atom or molecule into a charged particle. The two main ionization processes in space weather are photoionization and impact ionization, where a photon or particle collide with the neutral particle expelling an electron.

ionizing radiation is either particle or electromagnetic radiation that has enough energy to ionize an atom or molecule. For living systems, ionizing radiation is a health concern and can cause radiation sickness, cancers, or death. In controlled and limited doses, ionizing radiation (such as X-rays) can be used as medical diagnostics or treatment of some medical conditions.

ionosphere is the region of the upper atmosphere where appreciable amounts of long-lived plasma exist. The Earth's ionosphere is characterized by several layers (D, E, F) at different altitudes where the plasma is created and lost through different processes such as photoionization and charged particle impact ionization.

kinetic approach is one of the main methods used to model the space environment. It solves the physics and chemistry of the system at the individual particle level.

Kp is a three-hour geomagnetic disturbance index that describes the general state of geomagnetic activity. It is a semi-log scale from 0 (quiet) to 9 (disturbed).

L-shell is the distance in Earth radii (r_E) to the equatorial crossing point of a dipole magnetic field line.

low Earth orbit (LEO) is a closed orbit around the Earth that has apogee below about 2000 km and is used for many Earth-observing and communication satellites. The orbital period is about 90 minutes.

magnetic field describes the force due to a permanent magnetic or electric current on moving charged particles, electric currents, or other magnetized objects. It is a vector and contains energy.

magnetic reconnection is the process of magnetic energy conversion to particle kinetic energy that changes the magnetic topology or connectedness of the interacting field lines. It is the process that powers solar flares, coronal mass ejections, and geomagnetic storms and substorms.

magnetohydrodynamics (MHD) approach is one of the main methods used to model the space environment. It makes the simplifying assumption that the plasma acts as a fluid over large scales.

magnetosphere is the region around a magnetized moon or planet dominated by the intrinsic, induced, or remnant magnetic field of the object. Intrinsic magnetic fields are produced by magnetic dynamos, while induced magnetic fields are created by plasma interactions in the body's ionosphere. Remnant magnetic fields (such as on Mars) are magnetized portions of the solid body that remain after the magnetic dynamo of the body ceased. The outer boundary is the magnetopause.

medium Earth orbit (MEO) is a closed orbit around the Earth that has apogee above low Earth orbit (LEO) and inside geosynchronous orbit (GEO).

models and simulations are important tools for all of science and engineering and are used in space weather to describe the behavior of systems. Simulations can run a model as a function of time on a computer to help understand the system and make predictions.

photoionization is the process that ionizes or knocks off a bound electron from an atom or molecule by an energetic electromagnetic photon.

photosphere is the 6000 K visible surface of the Sun, often characterized by the presence of sunspots.

plasma is a state of matter where the atoms and molecules are ionized and therefore electrically charged and subject to electric and magnetic forces.

radiation effects on satellites are one of the main space weather impacts on operating spacecraft and include spacecraft charging, deep dielectric discharge, and single-event effects – all of which can damage or destroy satellite systems.

radio wave propagation is impacted as it traverses the ionosphere and can lead to refraction, reflection, absorption, and scattering of the signal. Observations of radio wave propagation through the ionosphere from radars, natural emissions, and man-made transmitters and satellites can be used to measure key parameters of the ionosphere such as density and velocity.

rems (radiation equivalents in man) are a unit of radiation dose that is scaled to account for the different impacts that different energy radiation can have on human tissue and organs. The average background radiation dose for Americans is about 0.3 rem per year.

satellite radio communication and navigation are modern space systems that use satellites to connect ground systems globally and through the global navigation satellite system (GNSS), providing position, time, and navigation services.

solar atmosphere consists of three layers: the photosphere, the chromosphere, and the corona.

solar cycle is the 11-year waxing and waning of the number of sunspots from solar maximum to minimum that corresponds to the changing magnetic structure and activity of the Sun.

solar flare is an explosive event on the Sun that releases large amounts of high-energy electromagnetic energy.

solar maximum is the phase of the solar cycle characterized by the largest number of sunspots on the Sun and is associated with the largest occurrence of space weather events such as active regions, flares, and coronal mass ejections. The solar magnetic field is highly structured.

solar minimum is the phase of the solar cycle characterized by the smallest number (or complete absence) of sunspots on the Sun. The solar magnetic field is at its most ordered and dipolar allowing the highest flux of galactic cosmic rays to travel into the inner heliosphere.

solar wind is the ionized solar plasma that constantly expands out supersonically from the Sun's corona pulling with it the solar magnetic field or interplanetary magnetic field (IMF). It consists primarily of hydrogen, a fraction of helium, and even smaller amounts of all the other heavier elements that make up the Sun.

space weather is the field of space science focused on understanding societal and technological impacts of the solar–terrestrial relationship.

Standard Solar Model is the basic physical model that describes the structure and processes that occur inside our Sun, explaining the observed dynamics and energy output.

stratosphere is the region of the Earth's atmosphere where the presence of the ozone layer creates a rising temperature with altitude. It is located above the troposphere and below the mesosphere, from about 10 km to 50 km above the surface.

substorms are geomagnetic disturbances characterized by explosive auroral displays in the midnight region, injection of energetic particles into the geosynchronous orbit outer radiation belt region, intensification of auroral electrojet currents, and their corresponding geomagnetic disturbances. They occur about 4 to 6 times per day.

supernovas and gamma ray bursts are two high-energy astrophysical events. Supernovas are the death throes of stars, while gamma ray bursts are thought to be generated by supernova explosions.

Systems Science is the interdisciplinary study of complex, coupled systems such as the Earth and its climate, and the Sun and its interaction with the Earth.

thermosphere is the neutral part of the upper atmosphere characterized by low densities and high temperatures. Low-Earth-orbiting spacecraft experience atmospheric drag due to the thermosphere.

transformer is an electrical device that can step up or down the voltage of electricity. Transformers are used in electric power transmission systems to change high-voltage and low-current electricity that travels long distances over transmission lines to useful low-voltage and high-current electricity for houses and buildings. They are susceptible to failure due to space weather induced GIC.

triangulation is the process of using trigonometry of triangles to determine the position and distance of an object using independent distance and angle determinations.

Van Allen radiation belts are torus- or donut-shaped regions of the Earth's dipole inner magnetosphere that have trapped energetic ions and electrons. The basic structure consists of an inner and outer belt separated by a "slot" region, and they were discovered by James Van Allen's experiment on Explorer 1 launched in 1958.

References and Further Reading

References

Akasofu, S.-I. (1964). The development of the auroral substorm. *Planet. Space Sci.* **12**, 273–282.

Ben-Ezra, Y., Pershin, Y. V., Kaplunovsky, Y. A., Vagner, I. D., and Wyder, P. (2000). *Magnetic Fractal Dimensionality of the Surface Discharge under Strong Magnetic Fields*. Technical report. Emek Hefer, Israel: Physics and Engineering Research Institute.

Carrington, R. C. (1860). Description of a singular appearance seen in the Sun on September 1, 1859. *Mon. Not. Roy. Astron. Soc.* **20**, 13–15.

Chesley, S. R. and Ward, S. N. (2006). A quantitative assessment of the human and economic hazard from impact-generated tsunami. *Nat. Haz.* **38**, 355–374.

Clarke, A. C. (1945a). V2 for ionosphere research? Letter to the editor. *Wireless World* **51**, 58.

Clarke, A. C. (1945b). Extra-terrestrial relays: Can rocket stations give world-wide radio coverage? *Wireless World* **51**, 305–308.

Dessler, A. J. (1967). Solar wind and interplanetary magnetic field. *Rev. Geophys.* **5**, 1–41.

Drake, S. (1957). *Discoveries and Opinions of Galileo*, translated with an Introduction and Notes. New York: Anchor Books.

Eastman, T. E. (2003). Historical review (pre-1980) of magnetospheric boundary layers and the low-latitude boundary layer. In P. Song et al. (eds.), *Earth's Low-Latitude Boundary Layer*, Geophysical Monograph Vol. 133, Washington, DC: American Geophysical Union. And supplementary table for "History of magnetospheric boundary layers" at www.plasmas.org/BL/

Eather, R. H. (1980). *Majestic Lights: The Aurora in Science, History, and the Arts.* Washington, DC: American Geophysical Union.

Eddy, J. A. (1976). The Maunder Minimum: The reign of Louis XIV appears to have been a time of real anomaly in the behavior of the sun. *Science*, **192**, 1189–1202.

Galilei, G. (1613). *Istoria e Dimostrazioni Intorno Alle Macchie Solari e Loro Accidenti*. Rome: Giacomo Mascardi.

Gamow, G. (1972). *Thirty Years That Shook Physics: The Story of Quantum Theory.* London: Heinemann Educational.

Gold, T. (1959). Motions in the magnetosphere of the Earth. *J. Geophys. Res.* **64**, 1219–1224.

Hess, W. N., ed. (1968). *The Radiation Belt and Magnetosphere*. Waltham, MA: Blaisdell.

Jokipii, J. R. and Thomas, B. (1981). Effects of drift on the transport of cosmic rays. IV. Modulation by a wavy interplanetary current sheet. *Astrophys. J.* **243**, 1115–1122.

Kivelson, M. G. and Russell, C. T., eds. (1995). *Introduction to Space Physics*. Cambridge: Cambridge University Press.

Leboeuf, J. N., Tajima, T., Kennel, C. F., and Dawson, J. M. (1978). Global simulation of the time-dependent magnetosphere. *Geophys. Res. Lett.* **5**, 609–612. doi:10.1029/GL005i007p00609.

Lightsey, R. (2001). *Systems Engineering Fundamentals*. Fort Belvoir, VA: Defense Acquisition University Press.

Loomis, E. (1869). Aurora borealis or polar light. *Harper's New Month. Mag.* **39**, 1–21.

Moldwin, M. B. (2006). Timeline of Solar–terrestrial Physics, https://space .engin.umich.edu/outreach/timeline-of-space-physics/ [accessed June 18, 2022]

Moldwin, M. B., Downward, L., Rassoul, H. K., Amin, R., and Anderson, R. R. (2002). A new model of the location of the plasmapause: CRRES results. *J. Geophys. Res.* **107**(A11), 1339–1355. doi:10.1029/2001JA009211.

Moore, T. E. (2020). The cosmic timeline of heliophysics: A declaration of significance. *Perspectives of Earth and Space Scientists*, 1, e2020CN000137. doi:10.1029/2020CN000137.

NASA Heliophysics Listing of Missions, https://science.nasa.gov/missions-page [accessed April 8, 2022]

National Space Weather Program (1995). *National Space Weather Program Strategic Plan*, FCM–P30–1995. Washington, DC: Office of the Federal Coordinator for Meteorological Services and Supporting Research.

Spinoza, B. (1963 [1677]). *Ethics*, trans. A. Boyle. Everyman's Library No. 481. London: Dent.

Watt, R. A. (1929). Weather and wireless. *Q.J.R. Meteorol. Soc.* **55**, 273–301. doi:10.1002/qj.49705523105.

Further Reading

Akasofu, S.-I. (1989). The dynamic aurora. *Sci. Am.* **260**, 90–97.

Alexander, P. (1992). History of solar coronal expansion studies. *Eos Trans. AGU* **73** (433), 438.

Appleton, E. V. and Barnett, M. A. F. (1925). Local reflection of wireless waves from the upper atmosphere. *Nature* **115**, 333.

Baker, D. N., Allen, J. H., Kanekal, S. G., and Reeves, G. D. (1998). Disturbed space environment may have been related to pager satellite failure. *Eos Trans. AGU* **79** (477), 482–483.

Baker, D. N. and Lanzerotti, L. J. (2016). Resource letter SW1: Space weather. *Am. J. Phys.* **84**, 166–180. doi:10.1119/1.4938403.

Barlow, W. H. (1849). On spontaneous electrical currents observed in the wires of the electric telegraph. *Philos. Trans. R. Soc. Lond.* **139**, 61–72.

Biermann, L. F. and Lust, R. (1958). The tails of comets. *Sci. Am.* **199**, 44–51.

Bolduc, L., (2002). GIC observations and studies in the hydro-Quebec power system. *J. Atmos. Sol.-Terr. Phys.* **64**, 1793.

Bone, N. (1996). *The Aurora: Sun–Earth Interactions*, 2nd edition. Chichester: John Wiley in association with Praxis Publishing.

Breit, M. A. and Tuve, G. (1925). A radio method of estimating the height of the conducting layer. *Nature* **116**, 357.

Burch, J. L. (2001). The fury of space storms. *Sci. Am.* **284**, 86–94.

Carlowicz, M. J. and Lopez, R. E. (2002). *Storms from the Sun: The Emerging Science of Space Weather*. Washington, DC: Joseph Henry Press.

Carrington, R. C. (1863). *Observations of the Spots on the Sun from November 9, 1853, to March 24, 1861, Made at Redhill*. London: Williams and Norgate.

Chapman, S. (1967). History of aurora and airglow. In B. M. McCormac (ed.), *Aurora and Airglow: Proceedings of the NATO Advanced Study Institute held at the University of Keele, Staffordshire, England, August 15–26, 1966*. New York: Reinhold, pp. 15–28.

Clark, S. (2007). *The Sun Kings: The Unexpected Tragedy of Richard Carrington and the Tale of How Modern Astronomy Began*. Princeton, NJ: Princeton University Press.

Clauer, C. R. and Siscoe, G. (2006). The great historical geomagnetic storm of 1859: A modern look. *Adv. Space Res.* **38**, 117.

Cliver, E. W. (1994a). Solar activity and geomagnetic storms: The first 40 years. *Eos Trans. AGU* **75** (569), 574–575.

Cliver, E. W. (1994b). Solar activity and geomagnetic storms: The corpuscular hypothesis. *Eos Trans. AGU* **75** (609), 612–613.

Cliver, E. W. (2006). The 1859 space weather event: Then and now. *Adv. Space Res.* **38**, 119. doi:10.1016/j.asr.2005.07.077.

Dessler, A. J. (1967). Solar wind and interplanetary magnetic field. *Rev. Geophys.* **5**, 1–41.

Dooling, D. (1995). Stormy weather in space. *IEEE Spectrum* **32**, 64–72. doi:10.1109/6.387145.

Drake, Stillman (1957). *Discoveries and Opinions of Galileo*, translated with an Introduction and Notes. New York: Anchor Books.

Dwivedi, B. N. and Phillips, K. J. H. (2001). The paradox of the Sun's hot corona. *Sci. Am.* **284**, 40–47.

Foukal, P. V. (1990). The variable sun. *Sci. Am.* **262**, 34–41.

Gaunt, C. T. (2016). Why space weather is relevant to electrical power systems. *Space Weather* **14**, 2–9. doi:10.1002/2015SW001306.

Gillmor, C. S. and Spreiter, J. R., eds. (1997). *Discovery of the Magnetosphere*, History of Geophysics Vol. 7. Washington, DC: American Geophysical Union.

Hewish, A. (1988). The interplanetary weather forecast. *New Scientist* **118**, 46–50.

Holzworth II, R. H. (1975). Folklore and the aurora. *Eos Trans. AGU* **56**, 686–688.

Jacchia, L. G. (1963). Variations in the Earth's upper atmosphere as revealed by satellite drag. *Rev. Mod. Phys.* **35**, 973.

Joselyn, J. A. (1992). The impact of solar flares and magnetic storms on humans. *Eos Trans. AGU* **73** (81), 84–85.

Kappenman, J. G. and Albertson, V. D. (1990). Bracing for the geomagnetic storms. *IEEE Spectrum* **27**, 27–33.

Kappenman, J. G. (1996). Geomagnetic storms and their impact on power systems. *IEEE Power Eng. Rev.* **16**, 5.

Kelvin, W. T. Lord (1892). Presidential address to the Royal Society. *Nature* **47**.

Knipp, D. J. (2017). Essential science for understanding risks from radiation for airline passengers and crews. *Space Weather* **15**, 549–552 (2017). doi:10.1002/2017SW001639.

Lanzerotti, L. J. (2017) Space weather: Historical and contemporary perspectives. *Space Sci. Rev.* **212**, 1253–1270. doi:10.1007/s11214-017-0408-y.

Lerner, E. J. (1995). Space weather. *Discover* **16**, 45–61.

Marconi, G. (1928). Radio communication. *Proc. IRE* **16**, 40.

Meadows, A. J. and Kennedy, J. E. (1981). The origin of solar–terrestrial studies. *Vist. Astron.* **25**, 419–426.

National Academies (2008). *Severe Space Weather Events – Understanding Societal and Economic Impacts: A Workshop Report*. Washington, DC: National Academies Press.

Odenwlad, S. (2007). Newspaper reporting of space weather: End of a golden age. *Space Weather* **5**, S11005. doi:10.1029/2007SW000344.

Parker, E. N. (1964). The solar wind. *Sci. Am.* **210**, 66–76.

Pisacane, V. L. (2016). *The Space Environment and Its Effects on Space Systems*. Reston, VA: AIAA.

Rishbeth, H. (2001). The centenary of solar–terrestrial physics. *J. Atmos. Sol. Terr. Phys.* **63**, 1883–1890.

Royal Academy of Engineering (2013). *Extreme Space Weather: Impacts on Engineered Systems and Infrastructure*. London: Royal Academy of Engineering.

Silverman, S. (1997). 19th century auroral observations reveal solar activity patterns. *Eos Trans. AGU* **78** (145), 149–150.

Siscoe, G. (2000). The space-weather enterprise: Past, present, and future. *J. Atmos. Sol. Terr. Phys.* **62**, 1223–1232.

Stern, D. P. (1989). A brief history of magnetospheric physics before the spaceflight era. *Rev. Geophys.* **27**, 103–114.

Stern, D. P. (1996). A brief history of magnetospheric physics during the space age. *Rev. Geophys.* **34**, 1–31.

Suess, S. T. and Tsurutani, B. T., eds. (1998). *From the Sun: Auroras, Magnetic Storms, Solar Flares, Cosmic Rays*. Washington, DC: American Geophysical Union.

Tsurutani, B. T., Gonzalez, W. D., Lakhina, G. S., and Alex, S. (2003). The extreme magnetic storm of 1–2 September 1859. *J. Geophys. Res.* **108**, 1268–1276. doi:10.1029/2002JA009504.

Van Allen, J. A. (1959). The geomagnetically trapped corpuscular radiation. *J. Geophys. Res.* **64**, 1683–1689.

Van Allen, J. A. (1975). Interplanetary particles and fields. *Sci. Am.* **233**, 161–173.

Index

Key concepts are indicated in **bold** in the index; page numbers that reference images are in *italic*.